A Brief Illustrated History of Machines and Mechanisms

HISTORY OF MECHANISM AND MACHINE SCIENCE

Volume 10

Series Editor

MARCO CECCARELLI

Aims and Scope of the Series

This book series aims to establish a well defined forum for Monographs and Proceedings on the History of Mechanism and Machine Science (MMS). The series publishes works that give an overview of the historical developments, from the earliest times up to and including the recent past, of MMS in all its technical aspects.

This technical approach is an essential characteristic of the series. By discussing technical details and formulations and even reformulating those in terms of modern formalisms the possibility is created not only to track the historical technical developments but also to use past experiences in technical teaching and research today. In order to do so, the emphasis must be on technical aspects rather than a purely historical focus, although the latter has its place too.

Furthermore, the series will consider the republication of out-of-print older works with English translation and comments.

The book series is intended to collect technical views on historical developments of the broad field of MMS in a unique frame that can be seen in its totality as an Encyclopaedia of the History of MMS but with the additional purpose of archiving and teaching the History of MMS. Therefore the book series is intended not only for researchers of the History of Engineering but also for professionals and students who are interested in obtaining a clear perspective of the past for their future technical works. The books will be written in general by engineers but not only for engineers.

Prospective authors and editors can contact the series editor, Professor M. Ceccarelli, about future publications within the series at:

LARM: Laboratory of Robotics and Mechatronics
DiMSAT – University of Cassino
Via Di Biasio 43, 03043 Cassino (Fr)
Italy
E-mail: ceccarelli@unicas.it

For other titles published in this series, go to
www.springer.com/series/7481

Emilio Bautista Paz · Marco Ceccarelli
Javier Echávarri Otero · José Luis Muñoz Sanz

A Brief Illustrated History of Machines and Mechanisms

Emilio Bautista Paz
Technical University of Madrid
Spain

Javier Echávarri Otero
Technical University of Madrid
Spain

Marco Ceccarelli
University of Cassino
Italy
ceccarelli@unicas.it

José Luis Muñoz Sanz
Technical University of Madrid
Spain

Additional material to this book can be downloaded from http://extra.springer.com.

This is a revised and updated translation of the original Spanish work "Breve Historia Ilustrada de las Maquinas" ETSII, Madrid, Spain, 2007.

ISBN 978-90-481-2511-1 e-ISBN 978-90-481-2512-8
DOI 10.1007/978-90-481-2512-8
Springer Dordrecht Heidelberg London New York

Library of Congress Control Number: 2010926023

Printed on acid-free paper

Springer is part of Springer Science+Business Media (www.springer.com)

Preface

Machines have always gone hand-in-hand with the cultural development of mankind throughout time. A book on the history of machines is nothing more than a specific way of bringing light to human events as a whole in order to highlight some significant milestones in the progress of knowledge by a complementary perspective into a general historical overview.

This book is the result of common efforts and interests by several scholars, teachers, and students on subjects that are connected with the theory of machines and mechanisms. In fact, in this book there is a certain teaching aim in addition to a general historical view that is more addressed to the achievements by "homo faber" than to those by "homo sapiens", since the proposed history survey has been developed with an engineering approach.

The brevity of the text added to the fact that the authors are probably not competent to tackle historical studies with the necessary rigor, means the content of the book is inevitably incomplete, but it nevertheless attempts to fulfil three basic aims:

First, it is hoped that this book may provide a stimulus to promote interest in the study of technical history within a mechanical engineering context. Few are the countries where anything significant is done in this area, which means there is a general lack of knowledge of this common cultural heritage. IFToMM, the International Federation for the Promotion of Mechanism and Machine Science (MMS), which has also collaborated in producing this book, is carrying out an important labour in this respect through the Permanent Commission on History of MMS, but more engineers need to be involved in historical studies. In addition, knowledge of the historical-technical developments of machines and mechanisms will lead to greater motivation to currently increase the efforts that are needed to obtain results that are useful for advancing technology and hence for society. The Hispanic cultural area is perhaps an example of this type of relative deprivation, particularly when compared with the English-speaking world. The Spanish Association of Mechanical Engineering (AEIM) starting from its former president Vicente Díaz, which has also collaborated in this book, is determined to promote this task in its sphere of influence. Not only the language of this book but also its general structure, with emphasis on the graphical descriptions, are aimed at attracting generations of mechanical engineering students to this field, who could use similar books as textbooks for optional subjects in their more completed technical formation.

A second aim is to pay a debt of gratitude to the often anonymous personalities, who throughout history have turned their ingenuity to the construction of mechanical systems that have contributed to the development of mankind. Indebtedness is also directed towards those who reflected on the fundamentals of machine designs and constructions to open up new horizons to civilisation. These men undoubtedly contributed more to mankind than many others whose names fill the pages of universal history, being they politicians, military men, or scientists. However, these men remain practically unknown. Remembering them is not only an act of justice; it is also, and maybe above all, the way to reveal a vital path to new generations of mechanical engineers and a stimulus to follow their example, with the pride of belonging to a tradition that is of unquestionable historical importance.

The final aim of this book is to stimulate a multidisciplinary thinking to fertilise the advance of knowledge with contributions from the different branches of human wisdom. There are too many stimuli in the present-day world that tend to pigeonhole the individual into ever more specialised fields and are therefore lacking in global vision. Mechanical engineering is also open to this risk and any attempt to open up new horizons will be more than welcome.

These ambitions are undoubtedly too many for such a small book, but they may give some idea of the enthusiasm that went into writing it.

Knowing their own history always strengthens a group's signs of identity. Building machinery and reflecting on the way it works has a long tradition in the past that continues with vigour in the present. Knowing the roots lends perspective to future actions by endowing them with a collective, continuous sense of development. If this book contributes to promoting this feeling, all the efforts will have been worthwhile.

This book would not have been possible without the help and support of many people. Those "authors in the shadows" have contributed ideas, images, and advice, which in one way or another, have led to the book's completion. Among the many names that should be mentioned are those colleagues from the Machine Engineering Division at Madrid Technical University (Pilar Lafont Morgado, Pilar Leal Wiña, Andrés Díaz Lantada, Héctor Lorenzo Yustos, Julio Muñoz García, and Juan Manuel Muñoz Guijosa) together with some other teachers and friends from other Spanish schools of engineering such as Felipe Montoya from University of Valladolid and even students as Raquel Bernardos. We should also like to thank Justo Nieto whose financial support through the "Foundation of the Valencian International University" has enabled the book to be published in a preliminary Spanish edition. The authors are also grateful to many colleagues within the IFToMM Permanent Commission on History of MMS who have helped them with comments and discussions during the last decade to become conscious that technical aspects of historical developments are worthwhile also for technical background and formation. Among the many colleagues from all around the world, the authors like to express gratitude to the last Chairmen of the Permanent Commission: Prof Teun Koetsier (from Amsterdam University), Prof Hong-Sen Yan (from Tainan University), and Prof Hanfried Kerle (from the Technical University of Braunschweig).

Apart from our gratitude to the persons closest to us, we must not forget that the pages of this book are full of machines and mechanisms that were thought out and drawn by brilliant minds that existed in the past, and without which there would be no raison d'être. The authors owe a debt of gratitude to all of them.

The Spanish authors would like to pay tribute to the memory of Professor Ignacio Medina. He was a fine example of the many people who have devoted their lives to science and the theory of machines and mechanisms. His teachings motivated both students and teachers in their study of this science. The figure in the cover represents a mechanism design for a pumping system by Francesco di Giorgio, as an example how an illustrated design can give a strong relevance of machine capabilities.

Finally, the authors are gratefully to their families whose patience and comprehension have permitted them to spend time and efforts on elaborating and completing this book.

April 2009 Madrid and Cassino

Contents

Introduction

The history of mankind is also, and maybe above all, the history of Technology. Culture in general and any culture in particular progresses with steps that are determined by its technology. Scientific knowledge itself requires a preliminary technological substratum for its progress. Studying a phenomenon is easy if the proper instrument is available to observe it. The astrolabe and the lens made astronomy possible in the same way that the microscope and the photographic camera served neurology. The connecting rod-crank mechanism has enabled all kinds of engines to be developed as well as automatic machinery for numerous activities.

In political and military matters, there is a presupposition in term of technological development. As Nebrija wrote about languages: "technology is always the companion of an empire". Economic history goes hand-in-hand with the technological development of goods and transport, which makes commercial trade possible.

From the great to the small, engineering produces and exploits technology existing at the time. Engineering demonstrates the worth of technological know-how and makes it available as a service to society by creating something useful and real. Each human group lives on a level that is strongly related to its engineering development.

Mechanical engineering is probably the predecessor of other forms of engineering and has persistently accompanied them up to our present age. The history of machines, therefore, embraces a very extensive period of the history of mankind.

On History Without Written Sources

Strictly speaking, history is based on written documents. This means that our knowledge of media, events, and personalities prior to writing is very blurred, as well as our knowledge of people with no written tradition. Cultures having a strict oral tradition enable the historical field to be widened somewhat, but the picture is not yet completely clear. Within our own Western culture, the stories of Homer and the Bible were transmitted orally for centuries before they were finally set down in written works. But in spite of this, they have not lost any relative historical value as well as is also the case with medieval ballads.

Material remains are the only source of historical information when written documents are not available, and these material remains basically provide information on the technology that was in the hands of the culture under study.

In order to describe chronological ages (Palaeolithic, Neolithic, Calcolithic, Bronze, Iron), technological criteria are used, likewise technological aspects are used as criteria, to classify large cultural groups (bell-shaped pot, vertical loom, Garamantian chariots, catamaran).

It is curious and significant for the aim of this book that the technology is the means of an analysis for outlining the historical human achievements.

Only for recent events (5,000 years as compared to hundreds of thousands years) have we other classification systems at our disposal. Only more recently, thanks to the references that cultures have developed through writing skills, we have data of technological aspects concerning the history of people that have populated most of the planet.

A large part of the historical remains that are available today can be considered within what is generally regarded as mechanical engineering. Thus, it seems appropriate to begin the history of machines with a description of those mechanical devices whose inventors or group of inventors remain unknown, and which are the foundations for subsequent developments.

On Written Documents

By considering their distribution in time and space, most written references that are related to machines can be found in treatises whose main object is not exclusively mechanical engineering. In books dealing with a variety of topics such as astronomy, agriculture, geometry, geography, architecture, or the art of warfare, mechanical systems are often described as mere curiosities or boasts of their ingenuity.

As far as we know, until recent times, specific systematic treatises on machines have been scarce. Apparently, only a few great civilisations have directed well thought attention to mechanical engineering.

But this occurrence probably does not correspond to reality. In the modern Western world, there are various reasons why present-day mechanical engineering is ignorant of large periods of its history.

Scarcely ever, in any culture, the “mechanical arts” have provided the prestige that has been awarded to other arts. Usually they have been just considered to be a “trade” and rarely have they been included in higher studies, so that they have been excluded from the effects of the intercultural osmosis that was brought through the circulation of books and their translation, and, therefore, excluded from a global flow of knowledge.

In addition, a technology has a commercial strategic value that is partially incompatible with the spread of knowledge and, of course, it is radically different from the value of science or art. The Theory of Relativity became meaningful when it was found to be useful for nuclear technology. Science tends to spread while technology tends to be confidential material.

Therefore, it is logical for political confrontation between cultures to have restricted the publication and dissemination of treatises on the successful construction of machines that could be used, for example, for war or navigation. It is logical that, taking account of it, we can find more written references that are related to machinery that was useful for agriculture or to automatons whose only use was recreational.

It is also logical that guilds that had managed to acquire a specific technology should accumulate their knowledge without publishing or spreading treatises about it. This would explain the continuing conjectures and opinions, because of a lack of written references from the period, on the mechanical systems that were used to build the pyramids and Gothic cathedrals in cultures that developed numerous treatises on other matters.

Nor should we forget the ethnocentrism that each great culture produces quite naturally and even automatically. Nor is the current western technological culture free of such ethnocentrism. A feeling of superiority due to our own achievements induces us to forget or at least not to fully appreciate the achievements of those around us. This can be automatically reflected in the available literature. Moreover, each culture has a predominant language by which to express itself with ease.

Written references that are in languages that are more remote in time and space become more difficult to consult in texts.

It is difficult, if not impossible, for present-day Western culture to approach Tibetan, Persian, or Ethiopian texts, even though they are written in languages that are close to those spoken today. More accessible languages, such as Sanskrit, Sumerian, and classical Greek usually entail serious difficulties for a typical machine scholar.

A manuscript in medieval Spanish may be scarcely intelligible for a present-day Spanish speaker. How machines and their parts are named changes radically with the passing of time. Each guild has always had its specific terminology and, in many cases, its own dialect. A typical case in Spain is the Pontevedra quarrymen's dialect, which is a language that is different from Galician.

Any history of machines will always encounter these difficulties and tend to give a wrong impression of the intellectual predominance of one culture over another.

Chinese pictographic writing is unusual for its concern with calligraphy. Thus, Chinese treatises are more accessible in our time than, for instance, the writings throughout time in the Indian subcontinent, where technology would logically have developed similarly.

On Cultural Influences on Design

Only recently has machine design become independent of concepts such as beauty, harmony, or magic that were part of the cultures that designed and used them.

Past machine designers' concerns with aesthetics may now appear to be superfluous when regarding many details. But harmony among the component parts was also a design criterion in the present meaning of the term. Harmony ensured better operation. The golden section was an indicator of perfection, even as related to mechanical performance.

All cultures have had abstract concepts of perfection that have influenced their mechanical designs. An animist, a Taoist, or a Buddhist applied design criteria that were affected by beliefs. This must be considered when examining what was produced.

When machines interacted with natural forces, religious, or magical feelings must also have had an influence on design activity. Present-day attitudes towards the environment reflect a similar reverential background towards some powerful avenging force that can bring punishment to those offending the ordered harmony of the Universe.

Since machines are mobile devices, animist and pantheistic cultures claimed they had individual souls, like the rivers, trees, and mountains. And for their creators and maybe their users, their behaviour would give a sensation of a living being, of something possessing a soul.

When reviewing the history of machines, it should be borne in mind that all these aspects that are linked to human nature have been influencing design for centuries.

On the Scope of this Book

Mechanical engineering embraces a very wide range of fields of technology. In order to limit the scope of this book to a manageable size, it only deals with devices that are made of moving parts.

Therefore, the contents cover limited areas of technical developments in also limited periods of time. The progressive knowledge of materials in prehistory does not enter into the contents of the book. A stone axe or a bronze weapon is not the object of this history; however a rotary hand-mill is. The existence of an oven is not taken into consideration but the mechanisms that are associated with its operation are. Mechanical manufacturing processes are only examined through the machines that are associated with those processes. A tool is only included if it is part of a machine tool.

Once the frame for the analysis has been limited, the chosen descriptive method is basically graphic, since machines have intrinsic properties for such a treatment. The book attempts nothing else than to give a wide-angle view of historical development without making an in-depth analysis of each of the presented examples. It does not claim to be encyclopaedic; it is simply a historical compendium.

Moreover, one of the peculiarities of this book is that it illustrates the attempt to outline the historical development of machines and mechanisms more from a technical point of view rather than a strictly history of science point of view, since the authors are mechanical engineers. They are not science historians and do not wish to be regarded as such but rather as experts who are interested and motivated to examine the most significant facts in their own area of knowledge of the theory of mechanisms. A full understanding of the historical development of technology also needs the help of experts in technical matters who can appreciate and reassess bygone achievements in the light of their own technical knowledge. Additional collaboration between science historians and technical experts is needed, as is currently the case in the field of industrial archaeology. Thus, this book is also an attempt to set out a technical approach in the field of the historical development of machines and mechanisms, but without too many technical details that will prevent its understanding from historical viewpoints.

At the beginning of each chapter, there is a global reference to the considered period through the most relevant facts, and the most significant treatises in the context of machine history. Following this introduction, each chapter contains a series of sections on the types of machines that are representative of the analyzed period together with illustrations that are related to the text and vice versa.

A fairly extensive bibliography will enable the reader to make a deeper historical analysis.

On the Contents of this Book

Producing a compendium on the history of machines is obviously not an easy task. This compendium has been structured with eight chapters that more or less correspond to historical periods, but they are centred on large cultural areas, which means there are inevitable overlaps in time. Each period has received a cultural heritage from the preceding one, and each culture is influenced by other cultural areas. However, the overall result leads to a reasonably coherent outline of Mankind's global development.

The first chapter, properly entitled "Anonymous Developments", shows examples of machine development in prehistoric times as well as the outcome of collective and popular resourcefulness, even in more recent times. Devices are discussed whose originators, are unknown to us for various reasons. The Neolithic revolution is also marked by anonymous machine development.

In historical periods, the two following chapters condense antiquity in the East and the West. If "History begins at Sumer", then that is where the history of machines begins. The great Mesopotamian and Indo-Ganges river cultures of south-east Asia or the plains of China, developed mechanical devices when they had acquired a knowledge of writing. The Chinese cultural area has been chosen to represent all of them for linguistic reasons that have been mentioned in previous sections. The continuity of this culture enables the chapter that is entitled "Chinese Inventions and Machines" to go from Antiquity up to the beginnings of the Modern Age.

Similarly, the third chapter on "Mechanical Engineering in Antiquity" has Greco-Roman culture as its central theme. Latin and Greek have been accessible languages for western people. This is why most references that were used for near eastern and north African cultures are taken from Hellenic and Roman authors who occupied and came into direct contact with extensive areas of the antique world. The influence of this antique world extended to the Middle Ages through Byzantium and Islam and it flowed into the Renaissance with the study authors like Vitruvius.

Chapter 4, on "Medieval Machines and Mechanisms", is mainly focussed on Arab authors, since Islam spread to the confines of the known world during the Middle Ages, and Arabic became the vehicle of culture in its area of influence.

Chapter 5, on "The Machine Renaissance", examines the enormous impulse of the time that was given to all spheres of knowledge, including machine study, mainly from Italy. Thus, a clear technological difference began to emerge between European culture and other existing cultures, which has continued up to the present.

Cultural dominance and Renaissance technology were not accompanied by the political dominance of those who created such culture due to the expansion of Turkey and the emerging national European states. But the technological impulse of the Renaissance first paved the way for great geographical discoveries and for building the European colonial empires that spread to a large part of the world.

Chapter 6 examines "Machines of the First Colonial Empires", which took advantage of the technological achievements of the Renaissance to exploit the available resources.

European colonial policy boosted industrial development by creating new, ever more scientifically-based technologies. Chapter 7 entitled "Machinery during the Industrial Revolution", reflects the progress achieved in this time. Technology, once again, went hand-in-hand with political dominance.

Recent history, particularly referring to machines, is deemed to be sufficiently known and therefore lacking in any of the "historical" interest that more remote periods of history have. On the other hand, the recent speed of technological development would make it difficult to attempt any systematic approach for our present period and would inevitably be ephemeral. The history of the machines dealt with in this book ends with the Industrial Revolution.

However, after the Industrial Revolution a singular event occurred in Europe which was quite unlike any occurrence in other cultures. This singular occurrence is reflected in Chapter 8, that is significantly entitled "A Vision on Machines", and it was mainly headed by the French enlightenment which, at the time, wielded an influence over all European countries.

This reflection obviously had prior origins, basically in the Renaissance, and gradually produced systematizations for personal machine study with ever more mathematical bases. This opened the way to subjects that have been included in higher study programmes. Our opinion was that this qualitative leap deserved a separate chapter.

The final part of this book contains a series of considerations on the future that may be deduced as a result of the history of machines.

Chapter 1
Anonymous Developments

Throughout the chapters of this book, the reader will be able to see that the preliminary information about machines and mechanisms is described as something already existing at the time of the treatises referring to that information. The information and the machinery itself already formed part of the cultural heritage of the community that was used to construct and use them. Only rarely, at the beginning of history, and then eventually with greater frequency, are machines presented as the results of conscious individual designs.

This anonymity becomes more explicit when a machine writer refers to cultures, which are different from his own, by attributing the invention to Chinese designers, Persians, or Celts, as being distant in time or space.

Machines and mechanisms are described as interesting curiosities, on the same level as customs, plants, or exotic animals. This has became clear with the appearance of ethnology as a structured science. Marcel Gauss, in his "Introduction to Ethnology" written as the basis for his lectures at the University of Paris, introduces technology and, in particular, mechanical technology, as one of the criteria for classification, together with economic, legal, or religious phenomena among others.

Similarly, the nineteenth century saw the appearance of societies that were devoted to regional studies of an ethnological nature, as a consequence of an interest in the visibly rapid disappearance of ancestral customs and ways of life, with the purpose of preserving them, at least in documented form. In the Spanish context, as in many other countries, societies of friends of the country were founded. Those societies also made frequent reference to popular machines and mechanisms. Many books were written on rural machinery with the same spirit but, unfortunately, most of the time they had only regional circulation. Emblematic examples are the Spanish books on Water Mills on Grand Canary Island by J.M. Díaz Rodríguez, and in Asturias, the book Hydraulic Devices by Gonzalo Morís.

Archaeology is an obvious source of knowledge for anonymous machine developments, but there are serious difficulties in recognising a finding of this kind that is not simply a single object but an ordered set of parts. For example, an isolated gearwheel may be misunderstood as a solar symbol, or a connecting rod,

E.B. Paz et al., *A Brief Illustrated History of Machines and Mechanisms*,
History of Mechanism and Machine Science 10, DOI 10.1007/978-90-481-2512-8_1,

for a schematic representation of a female god. Only if parts are discovered in connection with the rest of a mechanism, is it possible to recognise their mechanical function. In general, in archaeological remains it is extremely rare to find fully intact a mechanical device. This makes an appropriate interpretation of those remains very difficult, particularly when the concept and design of a machine is related to the relative movements of several components. Since current archaeology gives more importance to the whole context of finds than to an isolated object, there is yet an intrinsic difficulty in recognising several scattered and incomplete parts as the components of a mechanism.

Finally, the most important development concerning anonymous machines are related to biological forms. Nature has continuously solved mechanical problems in order to construct and maintain life and has demonstrated brilliant ingenuity and creativity in the design of highly efficient mechanisms. This short book would have a significant gap if it would not deal, at least succinctly, with this part of machine history.

Through the above-mentioned observations, it is evident that there is difficulty in finding proper sources for this chapter, and even more difficulty in ordering them as they come from such different fields as archaeology, general and regional ethnology, and biology. The authors are aware of the limitations and omissions in this respect.

On Machines Before Man

Mechanical design existed even before man appeared on the Earth. A human being itself is a complex machine. The Universe looks like a machine in motion, mainly with well-defined cycles that are subject to mechanical laws as was observed by the first men and the first civilisations as a "harmony of the heavenly bodies" and even influencing events and individuals. In a sample view, Biology can offer us examples of mechanical design, like the leg of an arthropod or a vertebrate (Fig. 1.1) whose design involved the solution of several mechanical engineering problems.

When fishes first emerged out of the sea into an air-filled environment, their fins had to be redesigned in order to adapt them to locomotion on land and to support their body weights, even by strengthening the materials of their structure (bone towards cartilage). The weight and locomotion considerably increased the actions within and towards the body when compared with the situation that had been practically weightless in the water. Even the tribological contact within the joints had to be redesigned with better lubrication. Thus, the leg mechanism became an effective means of locomotion over all types of terrain (Fig. 1.2), as well as being useful for to climbing or digging. Another leg redesign turned vertebrates into flyers when the basic leg design was modified to evolve into a wing (Fig. 1.3).

In addition, some vertebrates returned to the sea, by further modification of the legs in order to make them similar to the fins of their ancestor fishes, and in even more likely, they became closer to the amphibians (Fig. 1.4) that were then living on dry land.

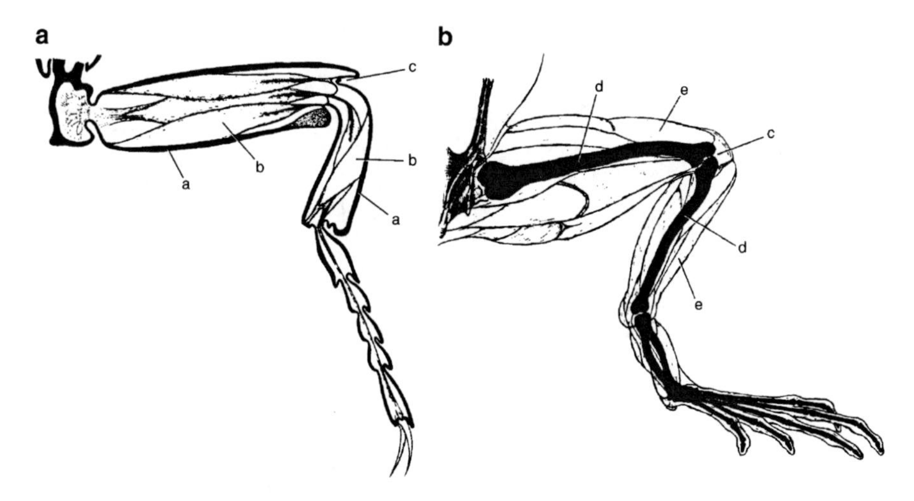

Fig. 1.1 Legs of a vertebrate and an arthropod [60]

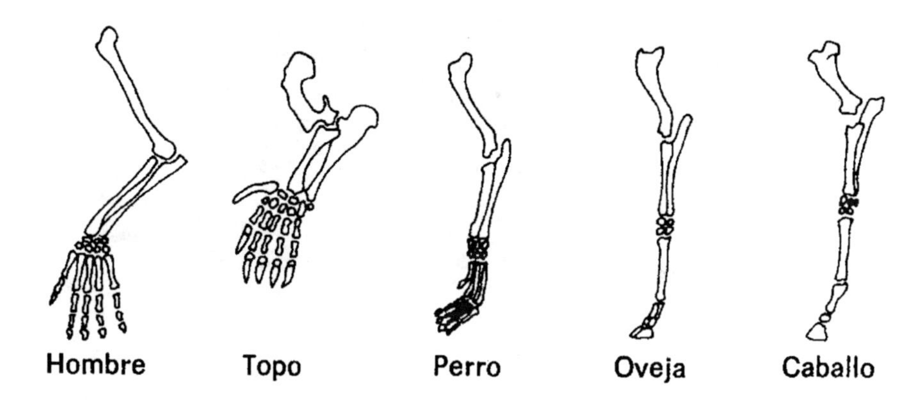

Fig. 1.2 Different leg designs [111]

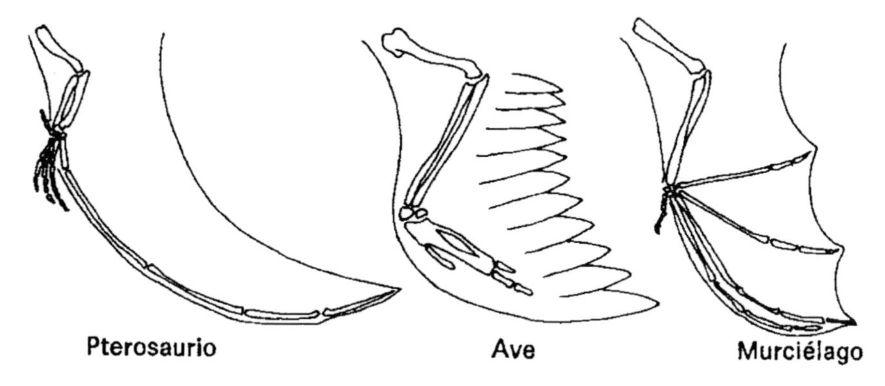

Fig. 1.3 The evolution of a leg into a wing [111]

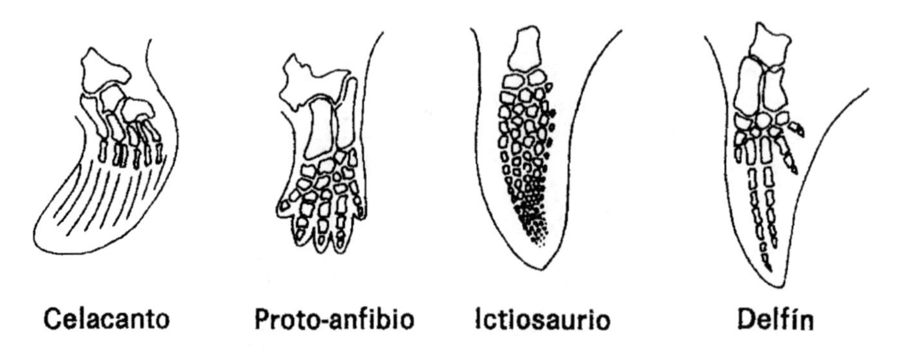

Fig. 1.4 Legs went back to the fin [111]

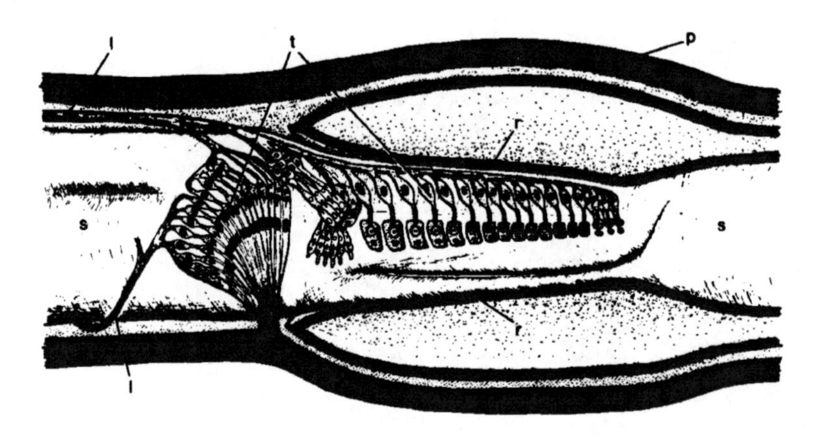

Fig. 1.5 Ear in a grasshopper's front leg [60]

Together with these original machines that made efficient use of the energy that is supplied by the muscles, Nature also designed what today might be called precision mechanisms. Sound analysis was performed by a system of vibrating sacs in a grasshopper's ear that is made with the aim to resonate by the different frequencies (Fig. 1.5). This is a very simple design as compared to other types of animal ears.

Another precision mechanism that is created by nature, is the moveable eye whose mechanical design with slight variations, is common to cephalopods and vertebrates (Fig. 1.6). It consists of a suitably lubricated spherical joint that allows the eye to be positioned rapidly and precisely in a proper desired direction. It also contains a system for filtering incidental light and a method of focussing by changing the lens curvature in mammals and by adjusting the lenses backwards and forwards in fish. These are only two of the mechanical features of the eye design.

Because of their exoskeleton structure, arthropods show the appearance of living machines and they are very interesting and inspiring from a mechanical engineering viewpoint. Beside the economy of means that is applied by Nature in this design, the mechanical perfection of a crustacean, an arachnid or an insect is

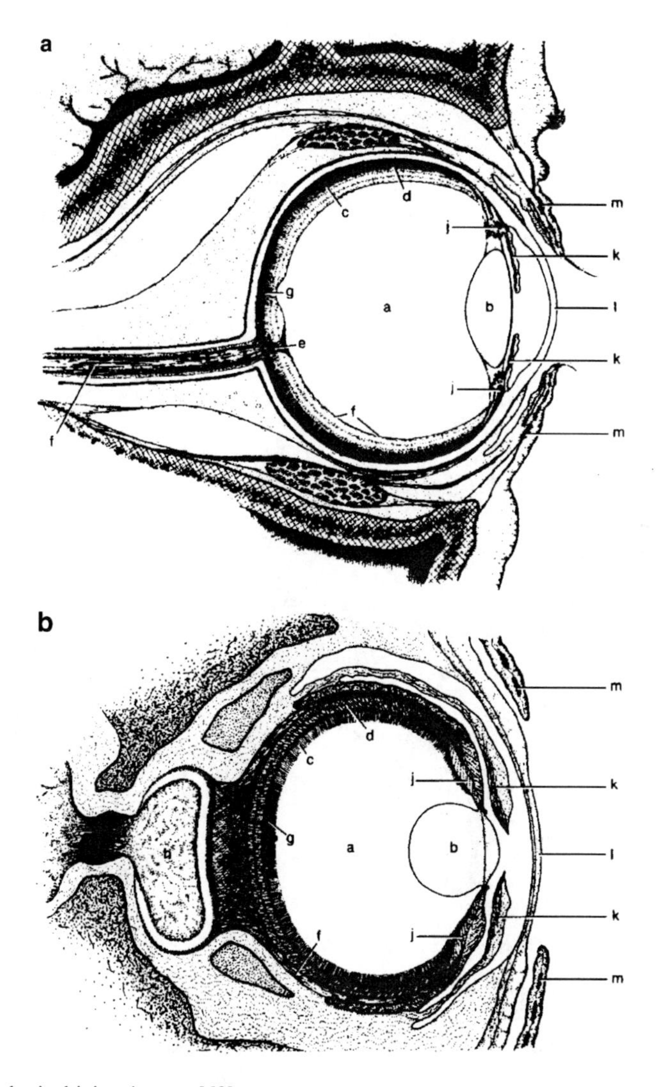

Fig. 1.6 Spherical joints in eyes [60]

amazing. With a similar anatomy, arachnids are designed (Fig. 1.7) in a way that their structures of locomotion are easily adapted to a large variety of environments like, for example, in scorpions, spiders, and mites. The precision by which the different moving parts are assembled together is really amazing. For example, there are one hundred in the case of a mite, and the overall size of the machinery leg is only few hundredths of a millimetre.

Animals themselves construct mechanical devices. It cannot be said that they make designs since they are already part of their genetic inheritance. But they properly adapt those designs to specific circumstances within which they use those constructions. This is the case of a spider's web or a weaver bird's nest.

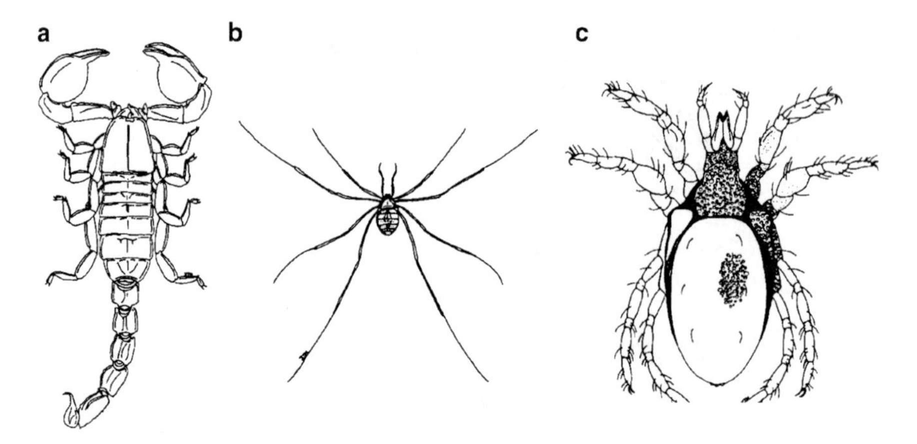

Fig. 1.7 Leg structures [77]: (**a**) a scorpion; (**b**) a spider; (**c**) a mite

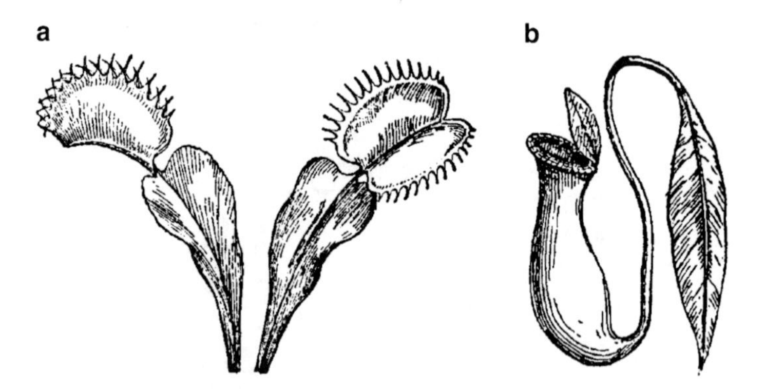

Fig. 1.8 Moving plants [37]: (**a**) Flycatchers; (**b**) Ascidia of Nepenthes

Some spiders make traps with moving parts which can be understood as real mechanisms. Carnivorous plants also have moving parts (Fig. 1.8a), that trap insects with movements that belong more properly to an animal rather than to a plant. This can be seen in the lid of the container in Fig. 1.8b that is used to protect the collected rainwater against evaporation.

Observing Nature must have inspired primitive men even until recent times, if we consider that ships copy the shape of ducks, including the head, and bird-like flying machines have been attempted.

On the Machines of Primitive Man

The concept of primitive man also includes a technological concept, which is better described as “a man that uses little developed technologies.” Cromagnon man’s mental capacity was similar to modern man’s, and present primitive cultures possess

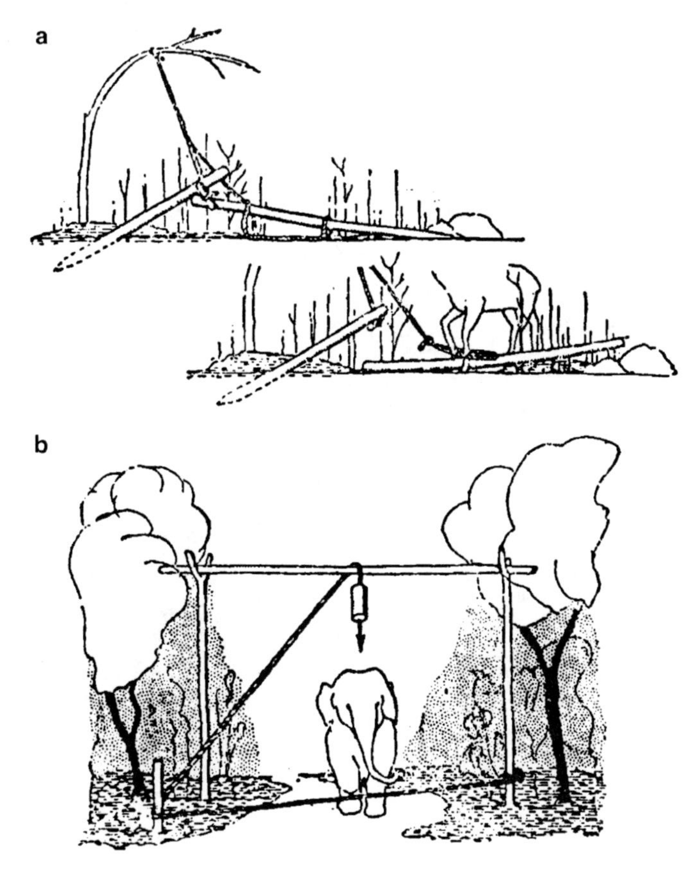

Fig. 1.9 Trap mechanisms [79]: (**a**) Snare for catching giraffes and other animals; (**b**) Elephant trap in East Africa

highly sophisticated languages and concepts for religion, cosmogony, and social views which can hardly be called primitive. Mastering technology can be understood as an arduous task that takes time and requires specific external circumstances. Oral transmission considerably limits technological developments and until the time when writing was invented man remained primitive. For a long period of time, man was a hunter-collector and continues like this in some current isolated cultures. Studying these cultures is of great help for understanding technological remains and techniques that were developed by mankind at his origins. Indeed, traps for hunting animals (Fig. 1.9) were the first machines built by man since they involved fast mechanisms that operated automatically as the animal passed through. The energy that is needed to operate these mechanisms ranges from gravity to energy supplied by vegetable springs that were also used in bows for arrows.

Observation of a spider's hunting process should have given the primitive hunter the idea to use nets to catch flying, land, and aquatic animals (Fig. 1.10). The trap was often completed with a closing mechanism or elastic catch that would allow passage in one direction only.

Fig. 1.10 River salmon trap of the Kwakiutl [79]

By looking at anthropological remains, man has used fire since remote times. In order to make fire by friction, he invented a machine that consists of a stick spinning fast that is driven by a bowstring (Fig. 1.11). Beside the mechanism's apparent simplicity, its successful operation involved the joint use of several devices, namely for converting translation to rotary movement, generating initial torque tension in the rope-pulley, multiplying the speed, supporting as a friction bearing, etc. It is curious that even today making fire is still linked to friction.

However, primitive cultures of south-east Asia use an ingenious mechanism (Fig. 1.12) that causes the tow to ignite by the air that is heated by adiabatic transformation similar to what happens in a modern diesel engine. The tinder is inserted into a cylinder which is plugged by the piston. A sudden blow to the piston shaft head causes a sudden decrease in the volume of air that is inside. This raises the temperature to a level that is required to ignite the tinder. A close fit is necessary between the diameters of the piston and cylinder. They must both also be made of suitable (wood) materials.

It is hard to say when those devices that present-day primitive man uses, were first used by the original primitive man. Those devices were manufactured with

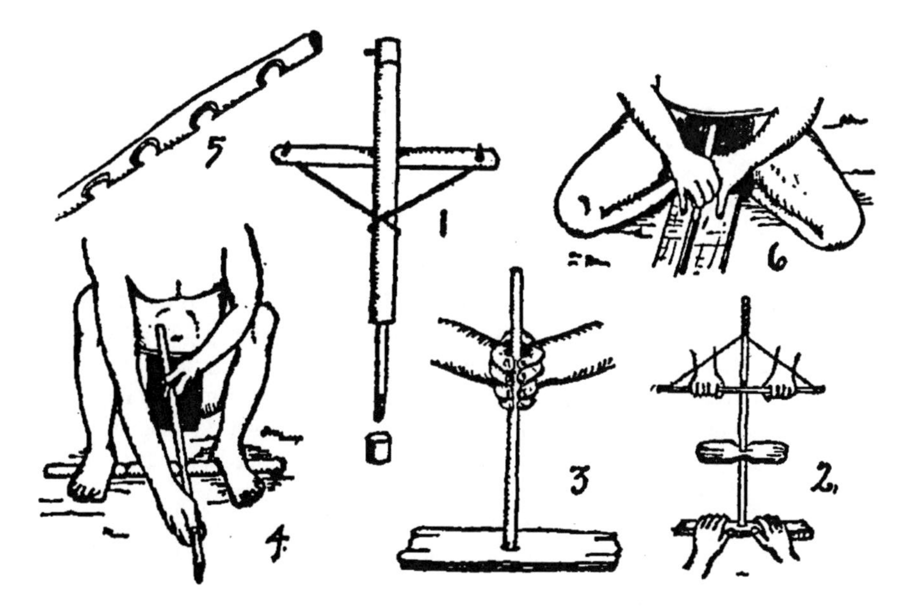

Fig. 1.11 Fire-making techniques [79]: (1) Yukaghir bow drill; (2) Pump movement drill; (3) Hand drill; (4) Saw for making fire; (5) Instrument from Queensland; (6) Melanesian fire plough method

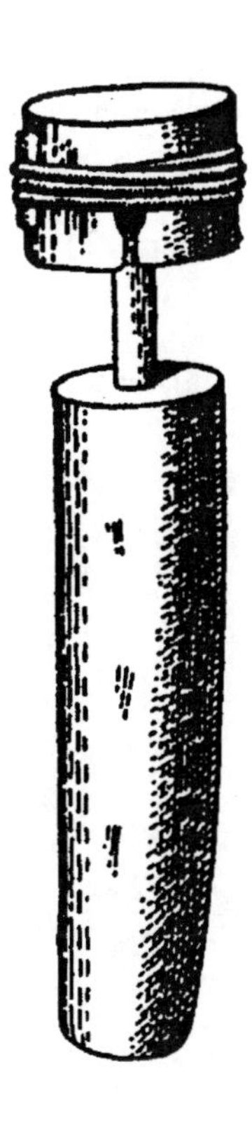

Fig. 1.12 Fire piston [81]

perishable vegetable materials, remains of which have not survived in Palaeolithic sites. For example, it may be assumed that the boomerang used by the Australian aborigines has been known since pre-history since there are European Neolithic cave-paintings that can be interpreted as such.

By pure logic, it is highly unlikely that many prehistoric mechanisms or machines made of wood will have survived at Palaeolithic sites. We should not think that the ingenuity of prehistoric man was less than that of present-day man belonging to primitive cultures, since the evolutions should be due to similar needs in similar surroundings. Thus, the mechanism that was designed to make fire had been used to drill both large and small holes (Fig. 1.13) and as a domestic device for churning milk to obtain butter (Fig. 1.14).

During the Neolithic revolution, over a few thousand years man domesticated animals and plants and he started a sedentary existence that enabled him (and maybe he was forced) to invent new machines. The wheel made its appearance with various uses, such as the potter's wheel (Fig. 1.15).

The development of agriculture made cereals an abundant food source and the human population rapidly increased. In order to grind the grain, more productive machines than the previous mortars were needed. In this case, even a wheel, made of stone, offered an appropriate solution as in the first hand mills (Fig. 1.16), where a spherical joint between the operating stick and concentric hole in the upper millstone ensured efficient operation. The apparent simplicity of the device reveals significant mechanical contributions when it is carefully analyzed.

Extensive agriculture is inseparable from irrigation. The first machine for raising water from wells or river beds was based on the use of a lever (Fig. 1.17), by using gravity as an aid to human effort.

Mechanisms similar to the lever have been known since the time of the hunter-collectors. The dart launcher in Fig. 1.18 gave a larger launching momentum than the provided by the hunter's arm. The fact that some of these launching mechanisms were made of bone or ivory has permitted them to survive in prehistoric sites.

Fig. 1.13 Mechanisms of fire making used as: (**a**) bow drilling machine in a reconstruction [119]; (**b**) fire drilling mechanism [79]

Fig. 1.14 Churn for butter drilling [79]

Fig. 1.15 Potter's wheel [79]

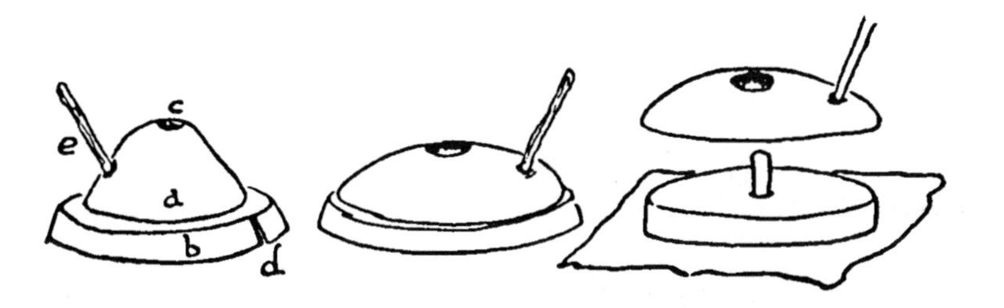

Fig. 1.16 Stone wheeled hand mills [40]: (**a**) Mill from Ifni, (**b**) Barley mill, Río de Oro, (**c**) Stone mill on hide, El Rehá (Río de Oro)

Fig. 1.17 Water extraction system with lever and counterweight (named as shadoof) [84]

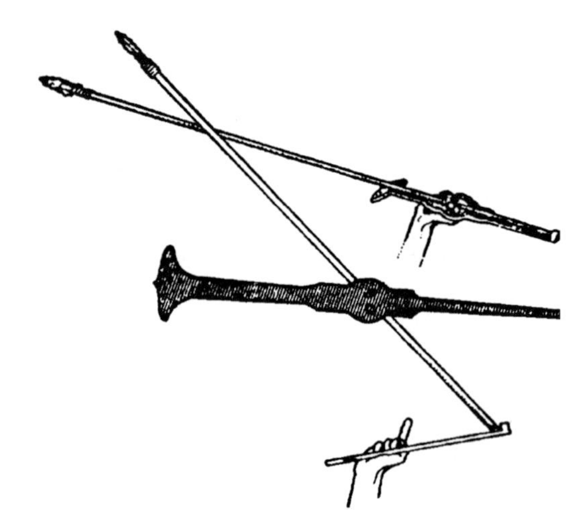

Fig. 1.18 Spear launching mechanisms [79]

Sometimes, a half-cylindrical bearing supported the dart with the moving end of the launching mechanisms, which reinforces the idea that the device was a precursor of well-developed mechanisms.

On Popular Machines

In rural environments machines are sometimes used whose designs show interesting variations of mechanisms that are already described in ancient treatises. It is difficult to know up to what point these rural craftsmen reinvented the machines or whether they started out from existing mechanisms. This problem of multiple origins within cultural dissemination is typical in ethnology when similar features are found in cultures that are far from each other.

However, it is worth noting that in their treatises, ancient authors wrote about machines that were used in their times as well as about other machines that were developed by well-known persons. Therefore, it seems not only fair to mention, even if only in a brief outline, the contributions of machines that were developed by the anonymous, inventive genius of ordinary people.

Very probably, the belt and pedal operated lathe appeared at the beginning of time and at least at the same time as the potter's wheel, with the same purpose of making containers, as shown in the example in Fig. 1.19. Since the parts turning in

Fig. 1.19 Conqueiro lathe [74]

the lathe were made of wood, it is unlikely that any archaeological remains will have survived, unlike ceramic containers. The machine itself, made with perishable materials, is unlikely to have resisted time. The same considerations can be made for popular textile machinery (Fig. 1.20).

Ordinary people's inventive genius was not only reflected in the conception of complete mechanisms, but also in the invention of ingenious mechanical parts. Figure 1.21 shows the pivot supporting an upright shaft, that was called, not surprisingly, an "egg", since it was a resource that the designer had at his disposal but with an appropriate choice of materials.

When mankind evolved to the Stone Age, several mechanisms became necessary to make metallurgy furnaces work. Ones of these were the bellows (Fig. 1.22a, b), whose operation is similar to the piston pump in Fig. 1.22c. Those devices depended of the invention of a valve system that permitted fluid to pass along in

Fig. 1.20 Spinning wheels [74]

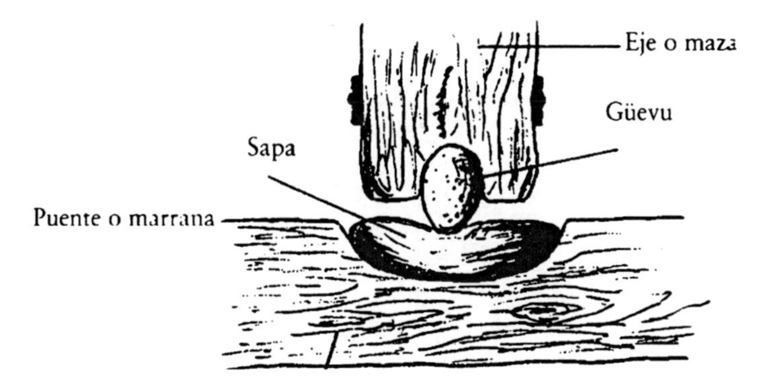

Fig. 1.21 Support system using an upright shaft bearing [84]

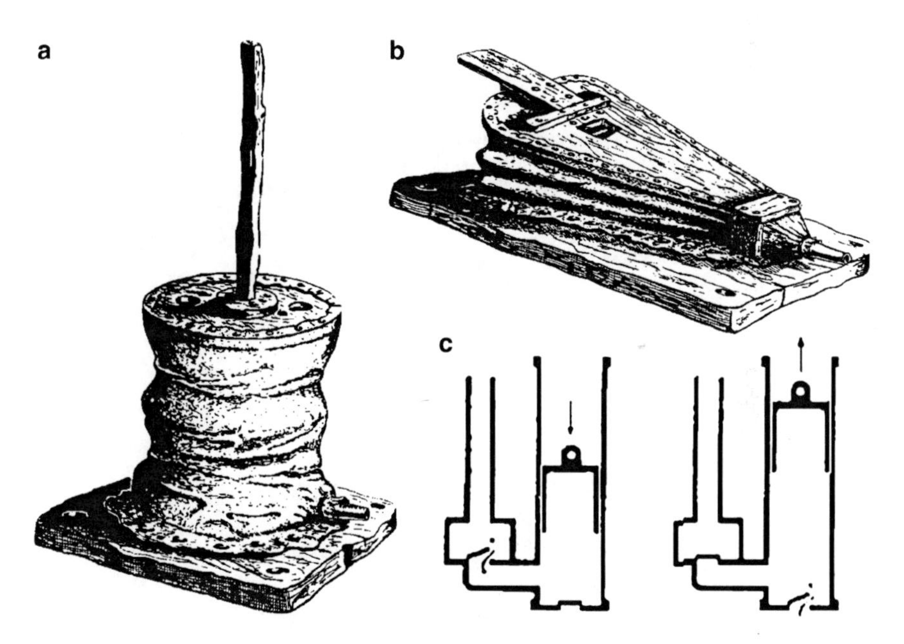

Fig. 1.22 Devices for air supplies: (**a**) a piston bellow and (**b**) planar bellow [84]; (**c**) a scheme for a piston pump, from "Ancient Greek Technology" by Elias Sfetsos

one direction only. The step from manual operation to the use of hydraulic energy as the driving force was only a question of time, but above all, it was due to a need for greater power.

But when hydraulic energy was already being used to supply furnaces with air, popular ingenuity came up with a new system that supplied the furnace

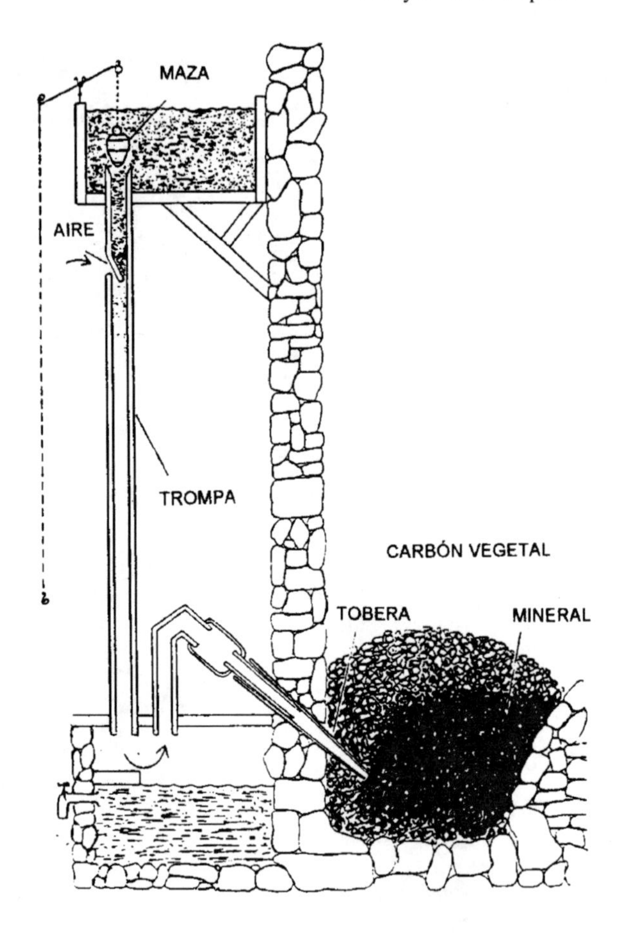

Fig. 1.23 Cross-section of a Catalan forge [84]

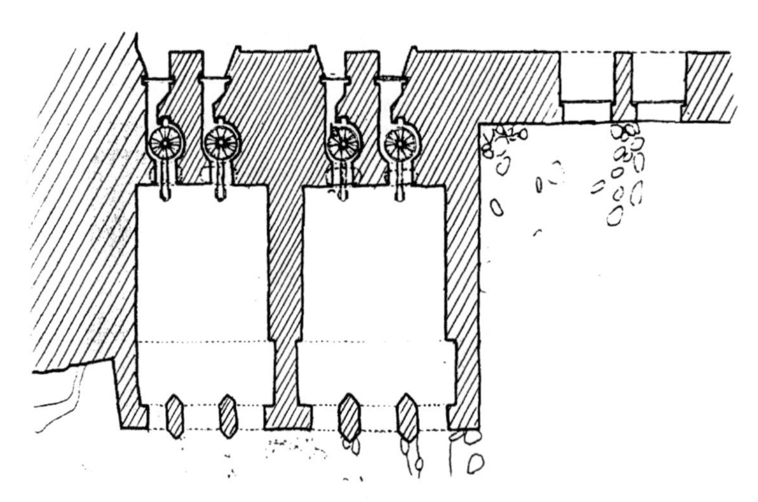

Fig. 1.24 A tide mill of thirteenth century [84]

with continuous air. An upright siphon (Fig. 1.23), through which the air flows, was the driving mechanism. In Compludo, which is a place close to Leon, there is a seventh century foundry that is still in use with this blast system, although nowadays it merely operates as a tourist attraction. It was not until the eighteenth century that Venturi explained the effect on which this system was based.

A further example of popular ingenuity is the use of tidal energy for operating mills. In the eleventh century, along the European Atlantic coast mills had been built by using the periodic changes in sea-levels as a power source (Fig. 1.24), since the tide is of relevant size. This new power source included changes to the mill design that were operated by river currents, as already described in Antiquity. The mill design in Fig. 1.24 was already working in Asturias in 1232, according to written documents.

Chapter 2
Chinese Inventions and Machines

For over 2,000 years Chinese society was pre-eminent in technological development. It was only at the beginning of the fifteenth century that it began to decline and was passed by Europe. Its technology began with agricultural, textile, and war machines; it was enhanced with hydraulic machines; and it was completed with the ingenious clocks and automatons that were built while the rest of the world was just waking up.

One of the first Chinese books on technology and craft dates from the Zhou Dynasty (770–221 BC): the "Kao Gong Ji" or "The Book of Diverse Arts", by an unknown author. It contains the knowledge gathered up to that time concerning astronomy, biology, mathematics, physics and engineering. Since then, Chinese technological development has constantly evolved with the help of explanations that were included in texts on manufacturing weapons, bells, dyeing processes, and irrigation devices.

Around the year 1040, Zeng Gongliang and Ding Du wrote the "Wu Jing Zong Yao". This book is a collection of the most important military techniques of the time and it includes 160 diagrams of machines. Among these diagrams are some catapults that have evolved considerably as compared to the first ancient models.

Regarding hydraulic engineering, the golden age can be identified somewhen between the tenth and fourteenth centuries. There are some books that particularly mark this evolution: the "Meng Xi Bi Tan" ("The Dream Swimming-pool", 1086) by Shen Kuo; the "Xin Yi Xiang Fa Yao" ("New Design for an Armillary Sphere and a Celestial Globe", 1089) by Su Song, and the "Nong Shu" ("Agricultural Treatise", 1313) by Wang Zhen. The last one is outstanding for the quality of its almost 300 diagrams and illustrations of the tools and machines for agriculture. Later, other titles can be found relating to a compendium of agricultural machinery such as the "Nung Cheng Chüan Shu" ("Complete Agricultural Treatise", 1628) by Hsü Luang-Chi.

These books deal at most with milling and water raising machinery, ranging from hand-operated mortars and crank-operated stone milling machines to crank winches for cranes. It was not until the seventeenth century that that all those machines were listed in the "Thien Kung Khau Wu" ("Exploring the Works of Nature", 1637).

E.B. Paz et al., *A Brief Illustrated History of Machines and Mechanisms*,
History of Mechanism and Machine Science 10, DOI 10.1007/978-90-481-2512-8_2,

In textile engineering, the book entitled "Keng Chih Thu" ("Drawings of Ploughing and Weaving", 1149) is an example of the great developments that was reached in silk manufacturing.

Chinese influence in surrounding countries was reflected at all levels, both scientific and cultural. An excellent example in the machinery field is the Japanese tea-serving automaton where the mechanics and assembly precision somehow overcame Chinese techniques.

With a few exceptions, there was no custom in Chinese culture of providing texts with drawings even though they dealt with technical matters. However, this changed due to the influence of European culture through the first Jesuit missionaries who tried not only to convert the Chinese people to Christianity but also to set Chinese culture towards European standards. Regarding machine techniques, European books dealt both with practice, like G. Böckler's "Theatrum Machinarum", and theory, like Guidobaldo del Monte's "Mechanicorum liber" and Galileo's "Le mecaniche". At the same time, they attempted to understand the Chinese machines with the same vision. Since the influence of the Jesuits, Chinese experts learnt to draw machines in order to show technical features.

This is why many ancient books have no drawings of described machines and illustrations are included from later editions that were produced in the seventeenth and eighteenth centuries.

On War Machines

One of the first known Chinese inventions is the catapult. Little is known about its origins but it was based on the principal of the lever (Fig. 2.1) that was already used for raising water from wells or channels into canals. The mechanism looks very like that of the Shadoof used by the Arabs (Chapter 4).

The Chinese army dominated its rivals for millennia thanks to its superiority in weaponry. It was able to expand its territories by means of rapid growth and control that were based on military supremacy and its geographical borders stretched from Tibet to the Pacific Ocean. Chinese technology not only evolved due to its engineers' brilliant minds but also because of a need to be superior to its neighbouring enemies.

The catapult was a basic element in ancient wars and the first known written reference to it is in the Mohist texts from the Period of the Warring States (fifth to third century BC). These texts describe the first version of what was later to be identified as a human powered catapult. These catapults were used to defend city walls by launching burning coals and logs or bottles of poison gas against the enemy.

Figure 2.2 illustrates the so-called "Whirlwind" catapult. This was a rotary catapult almost 2 m high that required the power of two men for an efficient launch based on the rotation of a horizontal shaft which drove an upright shaft and the projectile. For greater effectiveness, a series of catapults set in a row were used to launch more projectiles in less time.

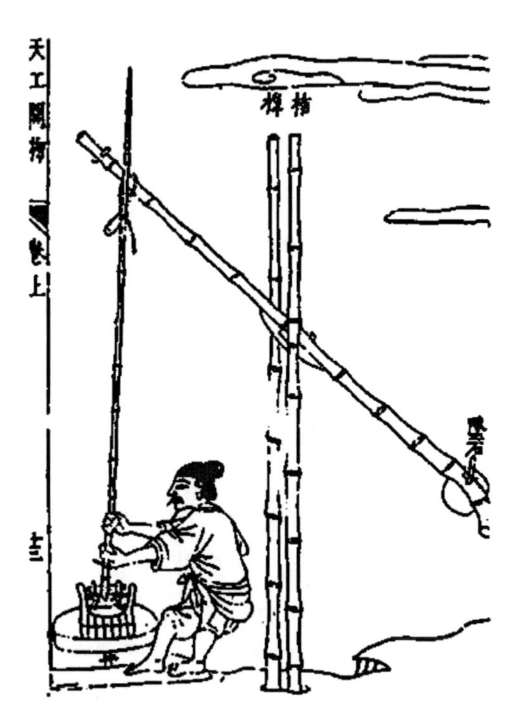

Fig. 2.1 Counterweighted lever used by farmers for obtaining water [86]

There are different types of catapults, namely the "trebuchets", tension catapults, and torsion catapults.

The "trebuchets" were based on the lever principle, the extremity of which was a counterweight to generate the force needed to launch the projectile over a long distance with a design similar to those shown in European sketches from the time of the Middle Ages (Chapter 4).

The so-called tension catapults stored energy by tensing a bow of wood. Torsion catapults took their energy from the spin of skeins of rope or tendons, while traction catapults were powered by human efforts. Examples are given in Fig. 2.2.

The illustrations in Fig. 2.3 show mobile catapults on wheels. The first illustration consists of an attack catapult which was used to launch or lift men. The operation power is obtained by human soldiers.

The second illustration in Fig. 2.3 shows the use of the kinetic energy that is accumulated by displacing the carriage. It was a heavy catapult that was developed at the same time as medium-range catapults known as "Hudun" (crouching tiger) or Xuanfeng light catapults. Large pebbles, animals, or grenades that exploded when launched were used as projectiles. It is believed that there were approximately 5,000 catapults in China around the year 1120 AD. This gives an idea of the use and need for this type of machine.

Another Chinese invention was the crossbow. It was used for the first time in 321 BC during the battle of Ma-Ling, although some scholars think that it was invented

a

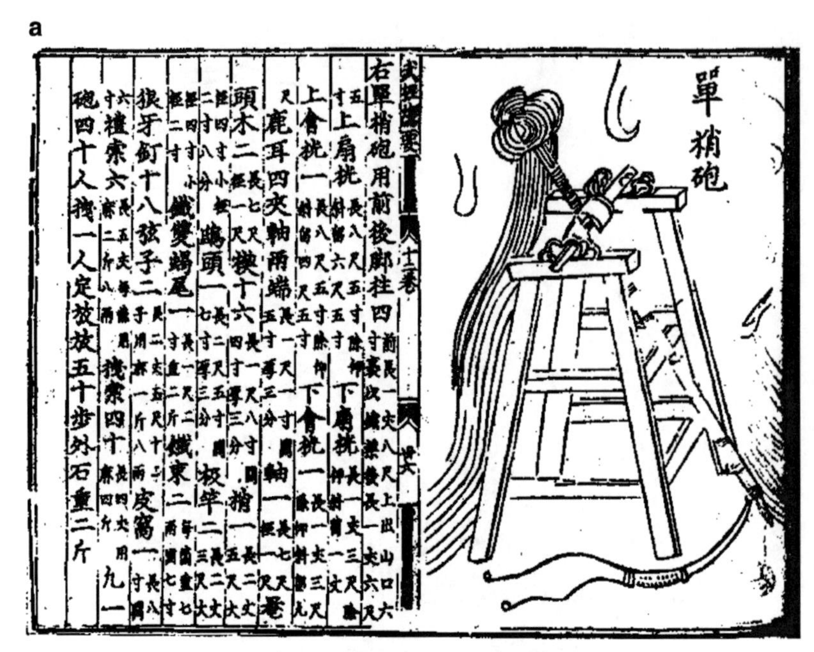

武經總要 卷十二

右單梢砲用前後脚柱四
上扇拢一
下扇拢一
上會拢一
下會拢一
鹿耳四夾軸兩端
軸一
頭木二
楔十六
梢一
鵝頭一
極竿二
鐵鑊鍋尾一
鐵束二
狼牙釘十八狡子二
皮窩一
檀索六
拽索四十九
砲四十人拽一人定放放五十步外石重二斤

b

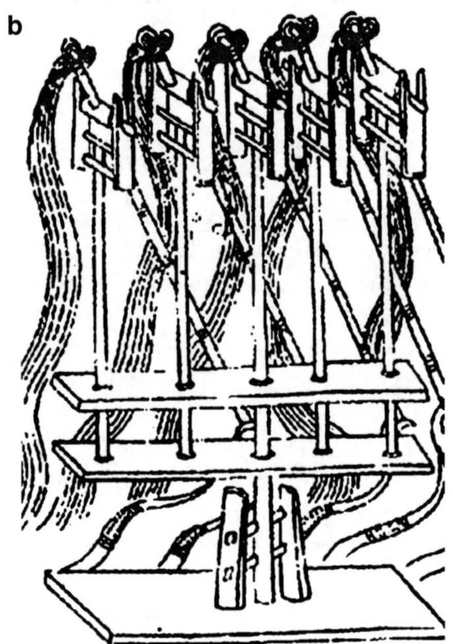

Fig. 2.2 A catapult and a battery of catapults [53]

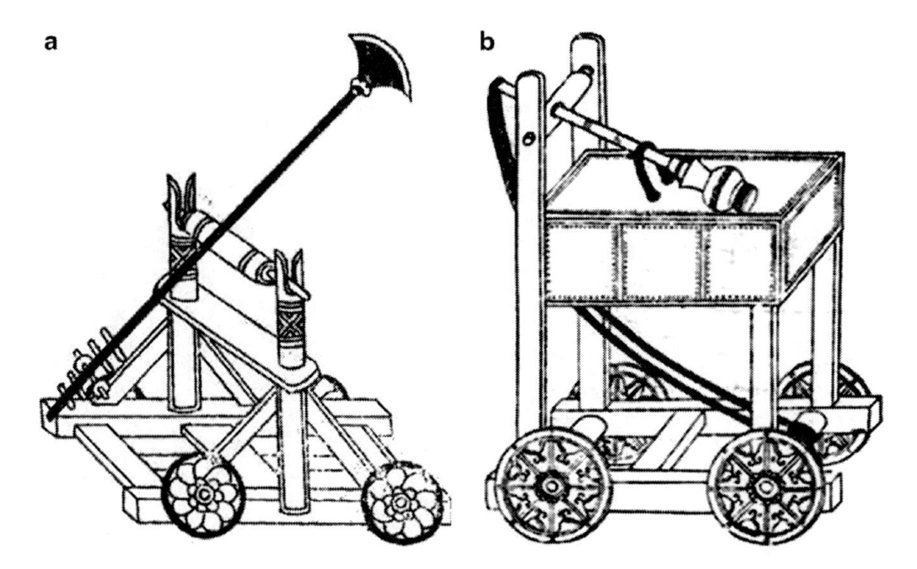

Fig. 2.3 An attack vehicle and a catapult [53]

in the seventh century BC. The crossbow evolved from the conventional bow and was widely used until gunpowder appeared on the scene, as another Chinese invention.

The crossbow consists of a horizontal bow and a release mechanism whose aim is to keep the arrow in place while the bow is tensed. When the release mechanism is triggered, the energy stored by the tension in a string is transmitted to the arrow which is launched with that energy. Usually, these crossbows had a longer reach than normal bows so that they were more effective than convention bows during battles.

Figure 2.4 shows a triple crossbow with a system for positioning it at a desired height for firing.

Summarizing, the described war machines involved the development of mechanisms for an efficient use of human energy. A lever enabled the multiplication of the launching speed. Both the heavy carriage and crossbow are based on the principle of energy accumulation (in inertial or elastic form according to the reported examples), which is released instantaneously on launching command.

On Textile Machinery

The manually operated distaff was the original concept for designing machines with three, five, and even ten spindles. Thus, efficiency of the manufacturing process increased so much that it was possible to supply all the population with clothes. This development gave an impulse to the textile industry and then to the country's

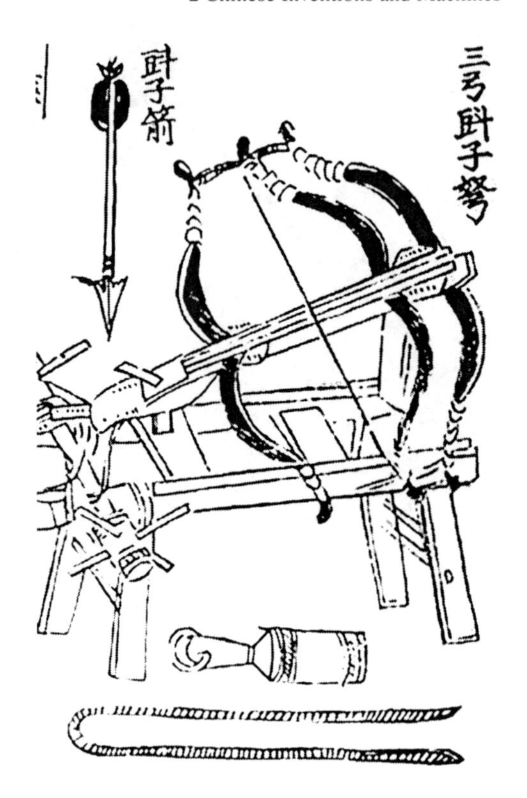

Fig. 2.4 Triple crossbow [86]

economy that lasted for more than four centuries during the Song (960–1279) and Yuan (1271–1365) dynasties. But since Antiquity, silk clothes were exported to the Roman Empire.

The textile machine in Fig. 2.5 was invented before the Christian era. Proof of its existence is due to its definition in a book entitled "Dictionary of Local Expressions" that was published in China in the year 15 BC. This machine was used mainly in the silk industry since it guides the silk threads to spools that are used by weavers. The mechanism is a multiple winder that is driven by a belt between the pulleys at each end. At one side, spindle shafts are operated by friction action. In addition, the shaft of a small pulley is operated with the aim of transmitting the movement to the shaft of the reels by means of another crossed-pulley system that permits a right-angle joint and, at the same time, a proper relationship between the rotations of the winder and spindles.

The design of this type of machine, together with similar spinning machines, can be compared to the first developments in textile machinery that appeared during the Industrial Revolution. The design in Fig. 2.5 shows a considerable similarity with Figs. 7.1 and 7.2 in Chapter 7, which illustrate machines that were supposed to be an "innovation", but 2,000 years later.

The powerful Chinese textile industry was forced to seek new energy sources as an alternative to human power, by giving important impetus to hydraulic power.

Fig. 2.5 A multiple winder

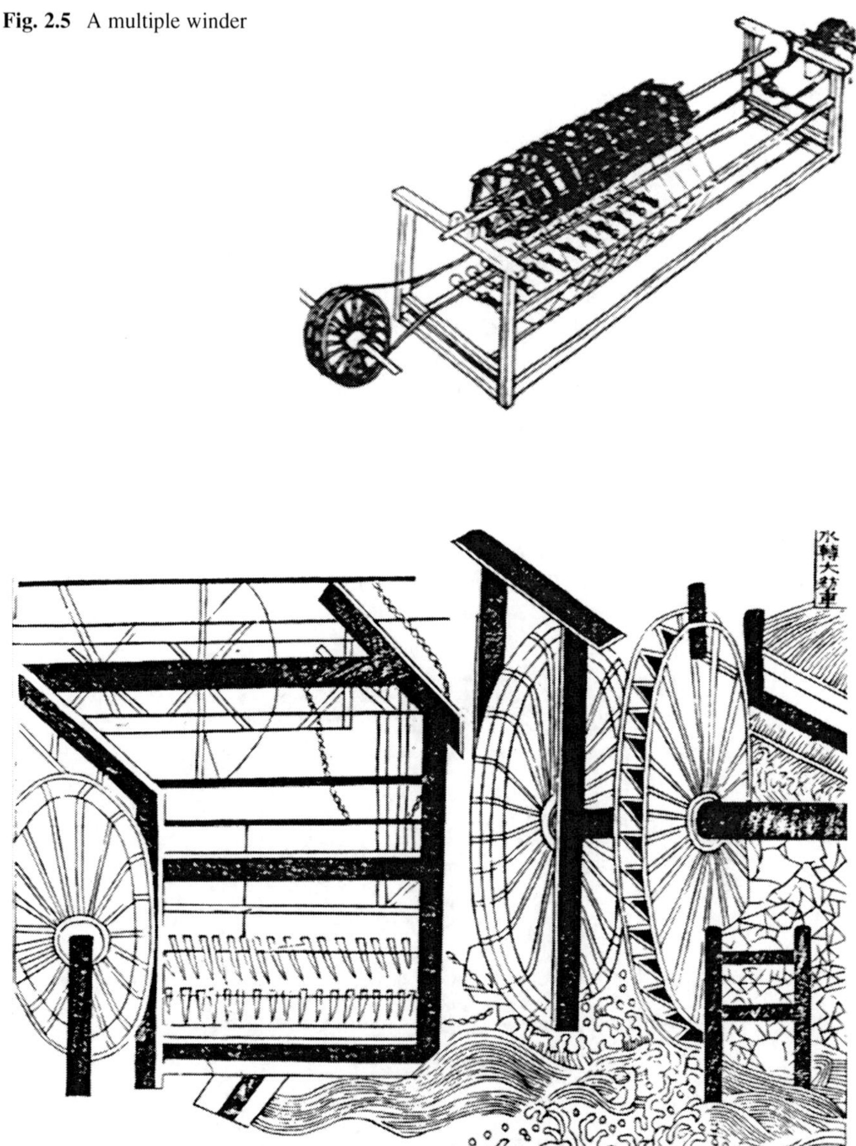

Fig. 2.6 Water mill power station for textile weaving (From the Nong Shu [86,141])

Figure 2.6 shows hydraulic power for operating a textile weaving machine as a means of increasing cloth production. By using hydraulic machines as power sources for the textile industry, China advanced with respect to other cultures that do not use the same process, until several centuries later.

On Hydraulic Machinery

In all cultures, hydraulic machinery is linked to irrigation for agriculture. The mechanism in Fig. 2.7 (possibly dating from the first century BC), consisted of a chain of buckets that is operated by a horizontal shaft through pedals attached to it. By observing the figure, it can be seen that the pedals were used by two men to rotate the shaft which is attached to a horizontal bar in the main frame of the machine. This action activates the wheel that moves the buckets which raise not only water but also sand or earth as required. The machine's performance depended very much on the perfection of its construction and the watertight solution by which water loss from leakage could be limited. Because of its efficiency, it is believed that this machine was capable of raising water to more than 4.5 m.

This invention spread throughout China thanks to its great utility and from there to the rest of the world centuries later. Since then, the variety of water wheels increased and different models were produced, made either of wood or bamboo (Fig. 2.8).

Figure 2.9 shows an important evolution. Besides a more refined and effective construction technique for transmitting movement, water energy has replaced

Fig. 2.7 Chain water elevator (From the Thien Kung Khai Wu [86])

Fig. 2.8 Bamboo water wheel (From the Nong Shu [86,141])

human power as the actuating source. This "water wheel for heavy weights", as it is called in Chinese books, made its first appearance in the book "The Nong Shu". On the other hand, the chain in Fig. 2.9 has several differences: instead of having bamboo buckets attached to the water wheel, they are fixed to each other by a rope, and the chain moves around the wheel where the bottom one is driven by the water and the upper one is used to realise the water.

Curiously, one of the most often-mentioned machines in ancient Chinese books is the drop-hammer water-driven machine. It was used to crush minerals or thrash seeds, and even in the metallurgy industry. The three figures (Figs. 2.10–2.12) show the changes from the twelfth to the seventeenth century. The first image of a hammer is as a man-operated tool (Fig. 2.10). But, because of its usefulness, it soon became necessary to operate it by hydraulic wheels. The first evolution can be observed in Fig. 2.10 where operation is not by hand but by foot. This gives a larger force for the operation and lets the operator's hands free to work. To produce the up and down movement, a hammer was attached to a perpendicular shaft that could rotate when supported by an upright structure.

Figures 2.11 and 2.12 show two models of water turbines that are used to operate a shaft with radial tongues that act as cam elements transmitting movement to

Fig. 2.9 Water wheel for heavy weights (From the Thien Kung Khai Wu [86])

several in-parallel hammers which, by falling down under gravity, give the desired blows in a synchronized, continuous operation.

Although the last two machines perform the same function, there is an important innovation differentiating them. While in Fig. 2.12 the waterwheel buckets are simple paddles that are installed perpendicular to the outer rings, the paddles in the hydraulic turbines in Fig. 2.11 are designed to optimise the hydraulic drive. With a difference of three centuries between Figs. 2.11 and 2.12, it is strange that the more modern one is less hydraulically efficient than the first one. Maybe the latter drawing was sketched simply to point out the mechanism itself rather than to show a detailed drawing of the water-wheel design. A machine that is referred to almost as often as the drop hammer in ancient Chinese books is the mill as a necessary instrument in everyday life for providing food.

In the mill in Fig. 2.13 there is a single millwheel that is operated by the lower horizontal turbine, which exploits the flow of water. It can be noted that two rings with paddles are used inside and outside the turbine in order to increase the efficiency with a simplified blade shape.

In order to ensure proper machine operation, a common practice was to fix the upright shaft to the ground rock by using cast iron as a cover for the hole.

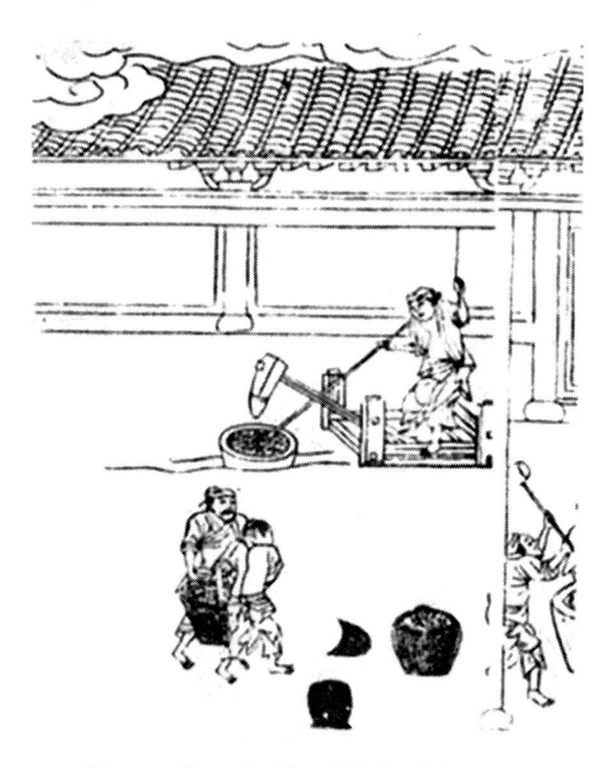

Fig. 2.10 Foot-operated hammer (From the Kêng Chih Thu [86])

This can be considered as proof of the considerable knowledge that the Chinese machinists had about materials and their properties in the thirteenth century.

Similarly to previous examples, the Chinese tried to multiply the output of a machine with a single actuation. An initial solution was the use of animal power, as shown in Fig. 2.14 for the case in a mill with eight millwheels that are operated by gearwheels. Some time after, the machine evolved and animals were replaced by a vertical turbine (Fig. 2.15).

The machine consisted of nine horizontal gearwheels and three vertical ones that are connected to the shaft of the hydraulic turbine. When the shaft turned, the vertical wheels also turned and generated the movement of the horizontal gearwheels so that the nine wheels turned at the same time. Compared to the mill in Fig. 2.13 a single hydraulic turbine operated nine millstones.

Another usual practice was to use a hydraulic turbine not only as a means of driving mills but also as a waterwheel to raise water and therefore to make the work twice as useful. It is known that around the year 1100 AD there were more than 250 tea mills. This gives an idea of how necessary this type of machine was and the reason behind the efforts to increase its output.

The knowledge of materials in Chinese society has already been mentioned, but among all the materials, metals were the most worked and studied. Specific machines were designed from the year 30 BC, like the one in Fig. 2.16. In this illustration,

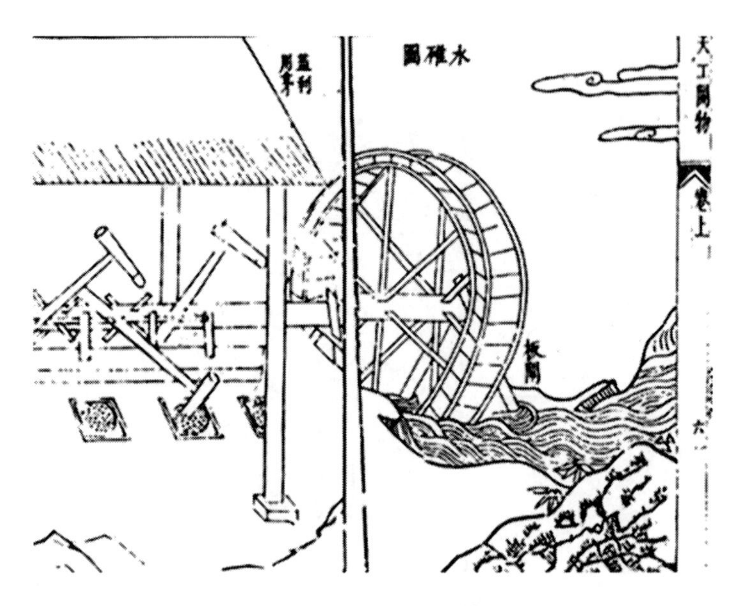

Fig. 2.11 Hydraulic drop hammer (From the Nong Shu [86,141])

Fig. 2.12 Hydraulic drop hammer (From Thien Kung Khai Wu [86])

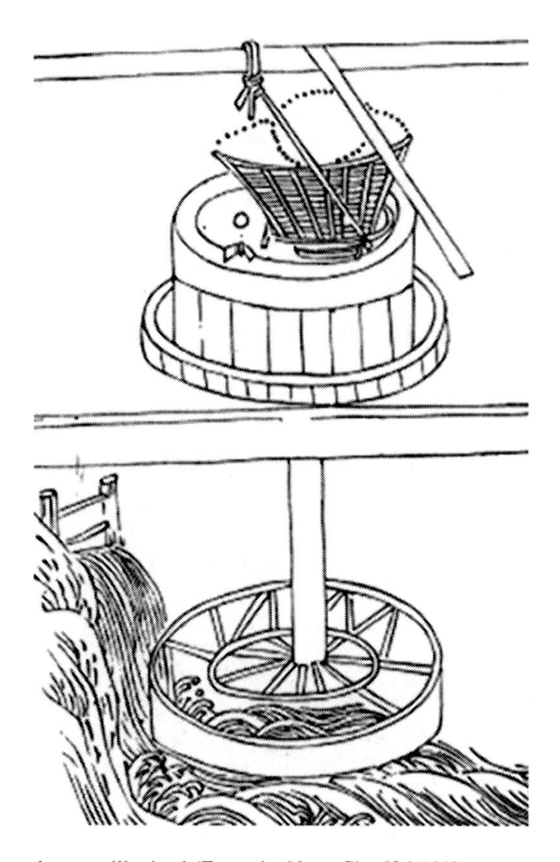

Fig. 2.13 Horizontal corn mill wheel (From the Nong Shu [86,141])

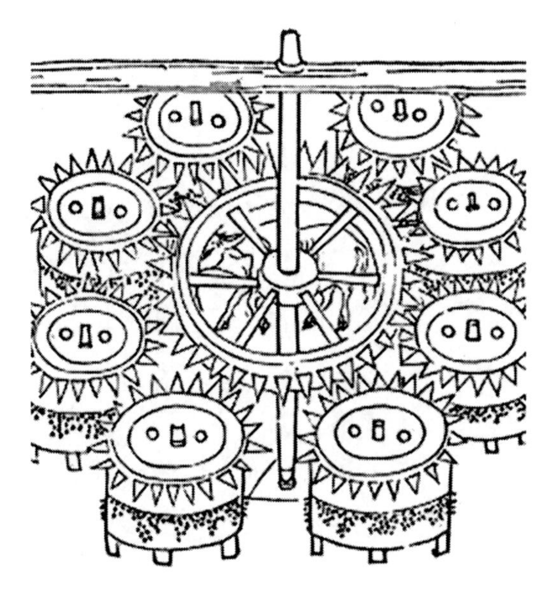

Fig. 2.14 Mill with eight wheels (From the Nong Shu [86,141])

Fig. 2.15 Nine mills (From the Thien Kung Khai Wu [86])

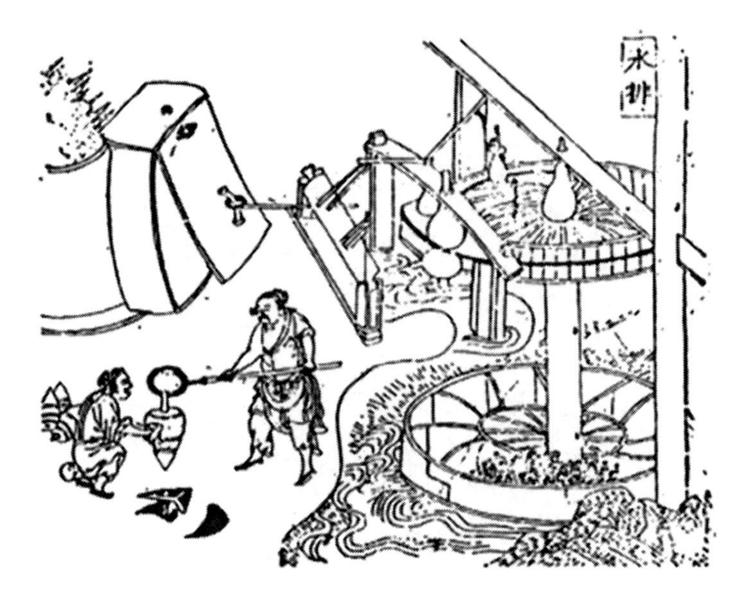

Fig. 2.16 Alternating motion generator (From Nong Shu [86,141])

the rotation of the waterwheel generates an alternating movement of bellows that maintain the fire by means of a beam operated by a horizontal cam. Remarkable are the double wheel and warped paddles that appear in the hydraulic turbine.

On Clocks and Automatons

One example of progress that was much beyond the comprehension of the rest of the world is "the south-pointing chariot" ("Chih nan chhê") that was built between 2600–1100 BC. The first written reference to it is found in a paragraph of "Sung Shu" ("History of the Sung Dynasty") by Shen Yueh in 500 AD: "The south-pointing chariot was first built by the Duke of Chou (first millennium before Christ) as a sure means of pointing the way home to those who had to return from a great distance beyond the borders".

It is also known that the workers who collected jade used it to get their orientation (Fig. 2.17). It was used for centuries in ceremonies of emperor worship since many emperors ordered its construction as a demonstration of their power. Its complex structure is shown in Figs. 2.18 and 2.19. It is composed of a series of gears

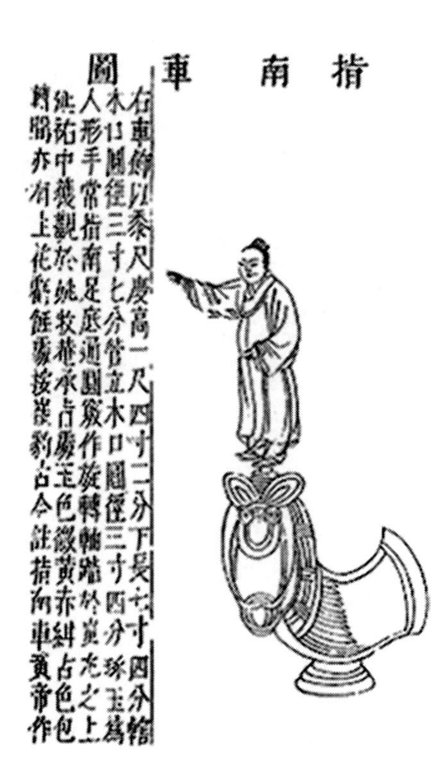

Fig. 2.17 Drawing of the "south-pointing chariot" (From the San Tshai Thu Hui [86])

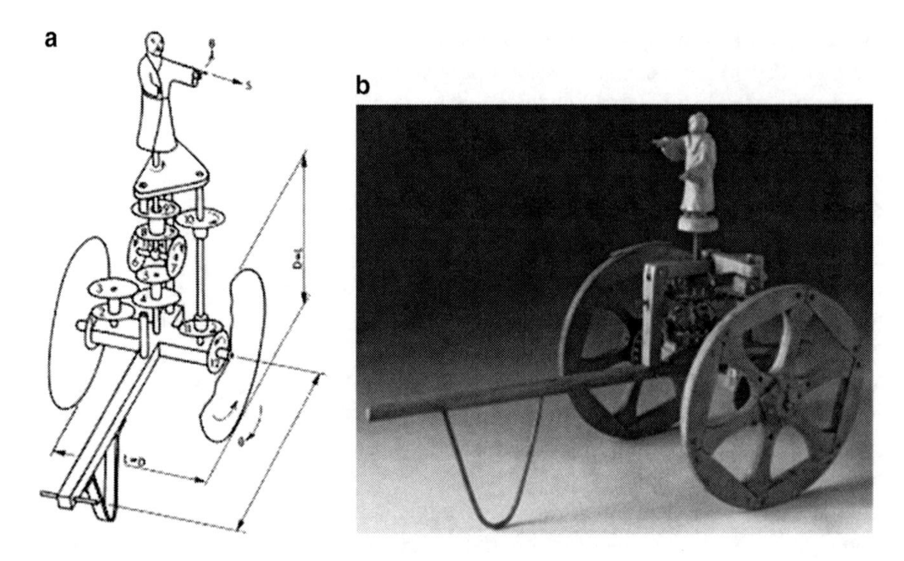

Fig. 2.18 South-pointing chariot: (a) a drawing, (b) a reproduction

Fig. 2.19 Gears of the south-pointing chariot

and gearwheels that always kept the figure's finger pointing south. The shape of these gears or the assembly has never been fully explained in any past writings, but it is believed they could have been as in Fig. 2.18a. This solution involves gear-wheels for each chariot wheel that is connected to a differential gear train, which locates the figure on top. Figure 2.19 shows details of the mechanical assembly.

Its operation works like a differential gear assembly. If the chariot moves along a straight line, the figure shaft does not to rotate. If the chariot turns, the wheels have different rotation speeds, which causes the upright shaft to rotate and thus the figure shaft rotates with it.

Much later, Su Song amazed the country by building an astronomical clock whose a reference is reported in his book, "Xin Yi Xiang Fa Yao". (AD 1089) (Fig. 2.20).

The clock took 4 years to be built and was completed in 1089 AD. The tower was between 10 and 12 m high. As a result of the mechanisms housed in its interior, various figurines appeared indicating the hours as well as beating drums and striking gongs while the movement of a celestial globe showed the stars and the constellations.

The clock's accuracy, less than a daily error of 100 of a second, was achieved by keeping the water in a tank at a constant level and this water operated the vertical wheel.

Figure 2.20 shows how the clock was made up of three levels. The upper level housed the armillary sphere (Fig. 2.21) representing the "the great circles of the heavens". This armillary sphere enabled astronomers to see the position of the stars around the Sun or the Earth. The sphere was composed of 12 rings in three layers with each of the rings marked on a different scale. Due to this layout, the positions of the 24 solar periods could be read directly and a planet or star could be located by looking through a tube.

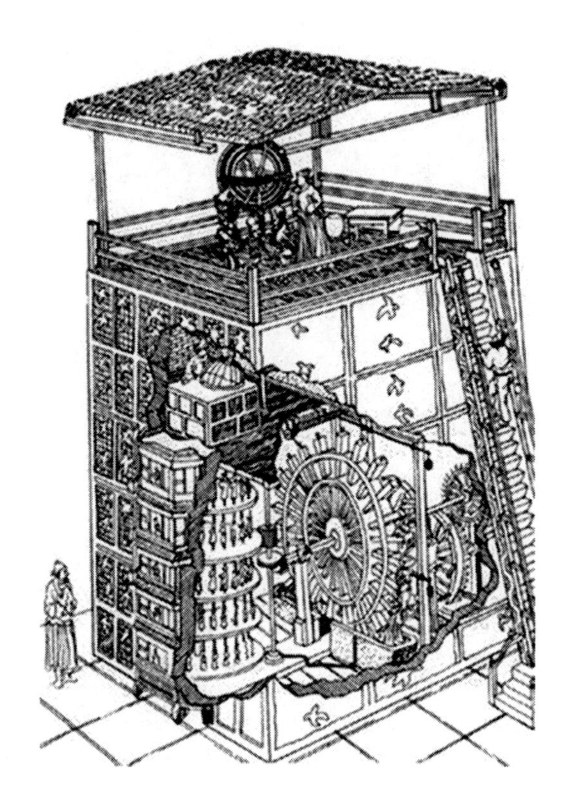

Fig. 2.20 Su Song's astronomical clock [128]

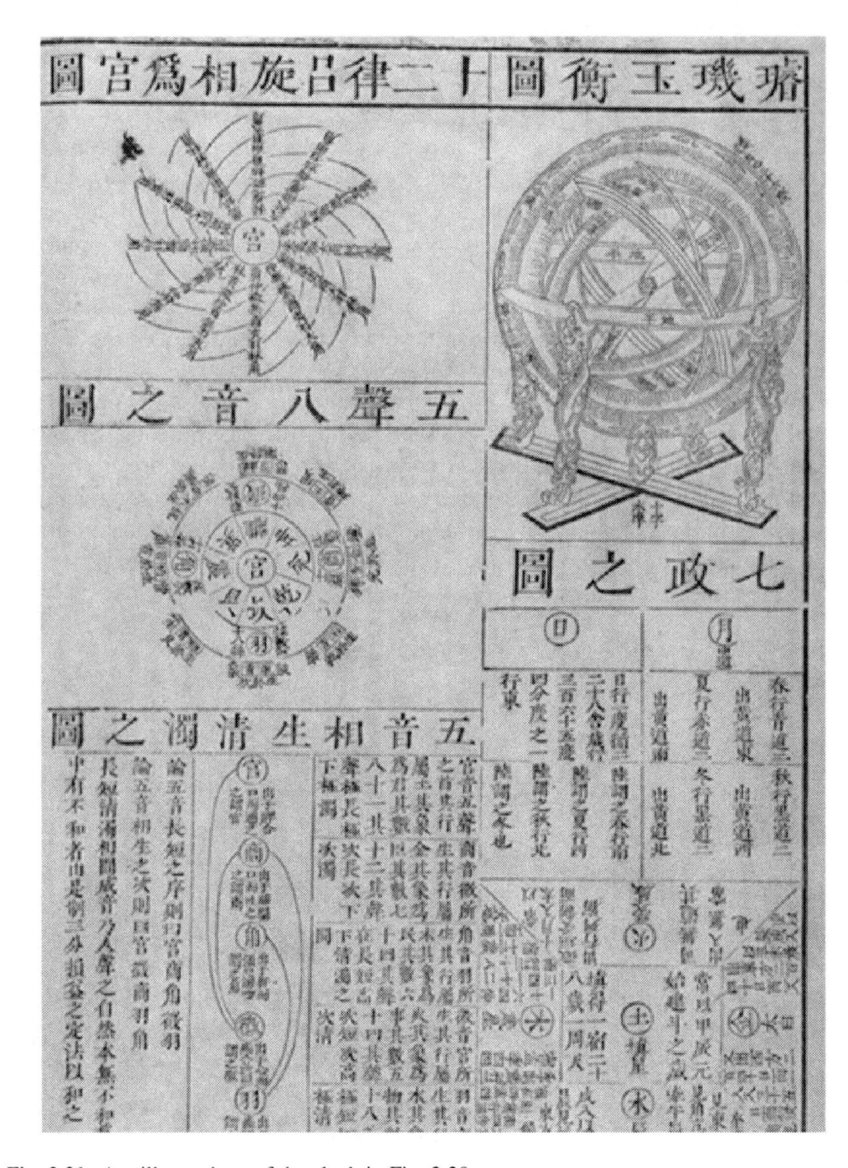

Fig. 2.21 Armillary sphere of the clock in Fig. 2.20

On the second or middle level there was a celestial orb showing the movements of the heavenly bodies which rotated fully in 1 day. The lower level had all the wooden mechanisms needed to make it work. Thanks to its precision, Chinese astronomers could not only observe the heavens but also make correct and exact measurements of the position of different heavenly bodies.

The astronomical clock only lasted 36 years before it was destroyed during the war in north China. Since then, various scale model replicas have been made

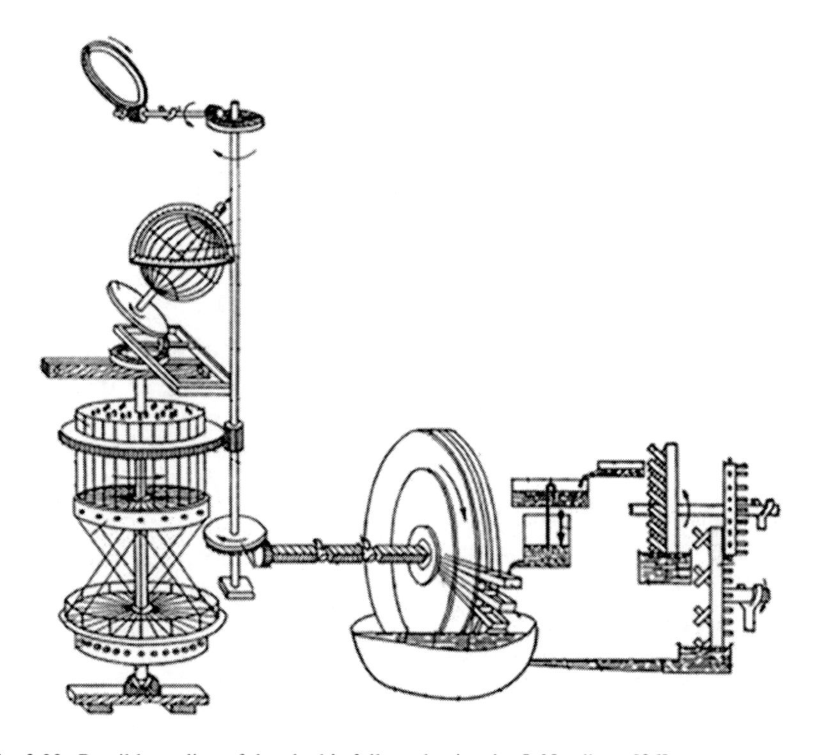

Fig. 2.22 Possible outline of the clock's full mechanism by J. Needham [86]

throughout the world in an attempt to unveil the secrets of its mechanisms which, even today, are still not fully known.

The outline in Fig. 2.22 shows the complex clockwork mechanism. It should be considered that all the gearwheels shown in the drawing had a specific number of teeth so as to be able to work with precision. Everything had to be carefully calculated so that the successive gears gave correct valid measurements that could be used by astronomers as proof of their ideas and theories concerning the heavenly bodies.

When the clock worked, the celestial globe and the armillary sphere turned to provide astronomers with the values of the variations of the Sun and other heavenly bodies throughout the day and seasons.

The most complicated part of the mechanics belonged to the water-operated wheel. Figure 2.23 shows part of its construction. The wheel was about 3.5 m in diameter; it had 72 spokes and 36 ladles which were attached to the outside ring of the wheel as can be seen in Fig. 2.23. The ladle was gradually filled with water, which poured into it at a particular speed so that the wheel turns under its weight. When the ladle was full, the counterweight was overcome and the wheel moved until the control system in the upper part made it to stop (the wheel was stopped at the next ladle) and the filling movement began again.

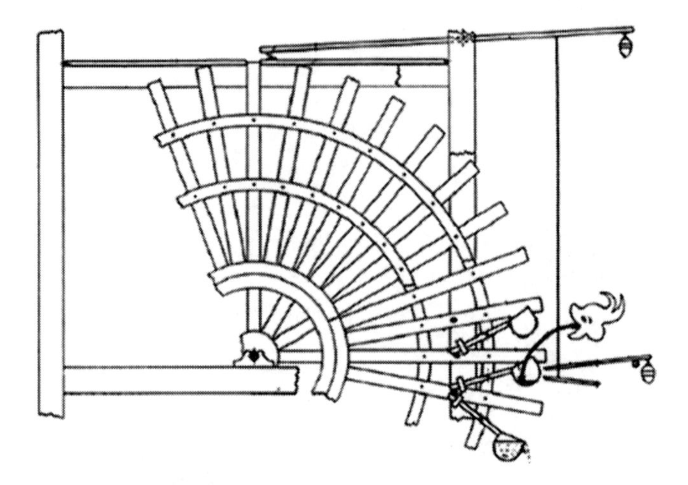

Fig. 2.23 Outline of the mechanism [86]

Once the wheel started its motion, this movement was transmitted to the drive mechanism that was made of a series of gear wheels. Su Song's book shows two different drive systems; namely, one that uses a main shaft that let the counting system rotate and another one with a worm gear that transmits the power to the rest of the system.

The time flow was gradually transformed into the movement of the armillary sphere, celestial orb, and the wooden pagoda that is composed of several interior rings. Accuracy in manufacturing and assembling the wheels was of vital importance.

The wheel of the drive system moved a shaft with six horizontal gearwheels that were hidden by a five-storey pagoda. Each floor had one or more doors. These horizontal circumferences housed different figures and mechanisms with sounds such as gongs or drums. When a row of doors was reached, the figures could be seen and the sounds of the musical instruments heard (Fig. 2.24).

In the book "Xin Yi Xiang Fa Yao", Su Song describes the clock by stating that it contains over 400 parts, including eight groups that work differently and were made of different mechanisms. Among the 63 drawings that are devoted to the clock, 47 are aimed at explaining how the mechanisms works.

In general, Chinese technology influenced both bordering and nearby countries. For the case of Japan, this influence led to the invention of automatons that are known as "Karakuri". They are mechanical devices for playing a joke, playing a trick, or surprising someone. These mechanisms were developed at the beginning of the Edo period (1603–1806) and were used both in religious and civil celebrations. Among all those automatons, that are undoubtedly precursors of modern robots, the best-known was Zashiki Karakuri: the tea server.

The tea server automaton shown in Fig. 2.25 had several different mechanisms that were combined with each other, whose job was to serve the tea. Once tension had been generated in the mainspring with a handle, a cup of tea was placed on the

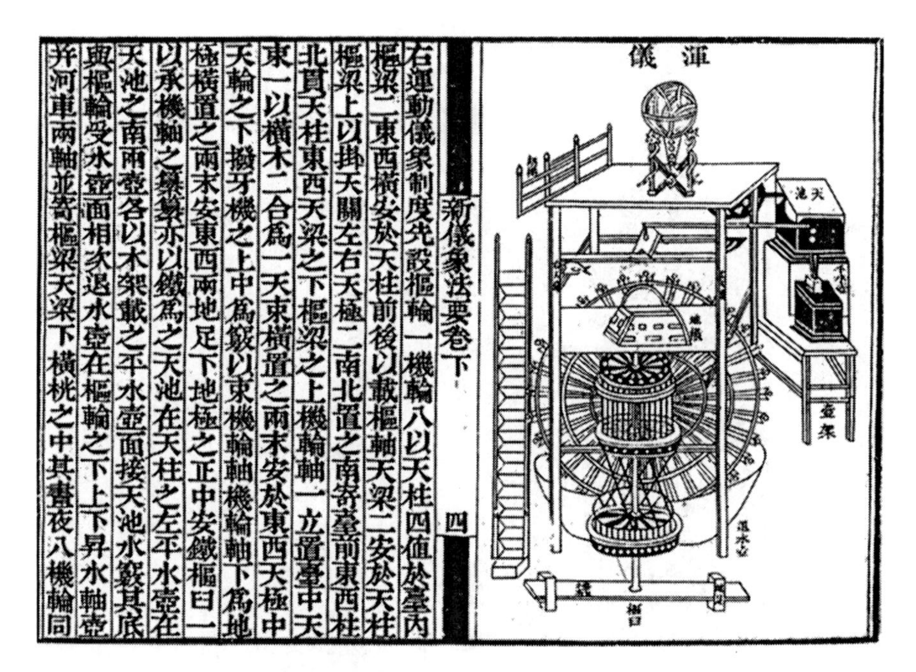

新儀象法要卷下　四

右運動儀象制度先設樞輪一機輪八以天柱四植於臺內
樞梁二東西橫安於天柱前後以載樞軸天梁二安於天柱
樞梁上以掛天關左右天極二南北置之南寄臺前東西柱
北貫天柱東西天梁之下樞梁之上機輪軸一立置臺中天
束一以橫木二合爲一天束橫置之兩末安於東西天極中
天輪之下撥牙機之上中爲轂以束機輪軸機輪軸下爲地
極橫置之兩末安東西兩地足下地極之正中安鐵樞臼一
以承機輪之纂纂亦以鐵爲之天池在天柱之左平水壺在
天池之南兩壺各以木架載之平水壺面接天池水竅其底
與樞輪受水壺面相次退水壺在樞輪之下上下昇水軸壺
并河車兩軸並寄樞梁天梁下橫桄之中其晝夜八機輪同

Fig. 2.24 Astronomical clock (From Xin Yi Xiang Fa Yao [114])

tray held by the doll which moved forward in a straight line until the cup was raised. Then it stopped and waited until the cup was once again put in its place. When pressure was once again felt on the tray, it did a half-turn and returned to whence it came. Moreover, as it walked away, it bowed its head deferentially up and down.

The way the automaton worked is described in the drawings in Figs. 2.26–2.28 that schematically interpret the description of Fig. 2.25.

Referring to Fig. 2.26a, when the tray is empty a spring causes a nail to be inserted between the teeth of the wheel, stopping its movement and making the automaton stop. The instant when a cup of tea is placed on the tray, the force of the spring is overcome and the nail rises by letting the automaton start again.

As the automaton moves, a cam mechanism (Fig. 2.26b) operates the bowing of the head, which rotates about the shaft that is indicated by a needle coming out of the neck. A spring ensures contact between the cam and the rod operating the load.

In the mechanism shown in Fig. 2.27, an eccentric rotating disc moves the fork that goes from the disc to the tea server's foot. This actuates the rocking motion of the automaton's foot and the backward and forward motion for the automaton walking. The eccentric discs of each foot are placed at 180° so that the movements of the feet alternate in moving forward so that the automaton walks without losing its balance.

The turning movement is obtained by a mechanism that is shown in Fig. 2.28a. It can be noted that the whole mechanism is in the lower part of the automaton. In order to keep the wooden lever upright, a spring is connected to the lever by a

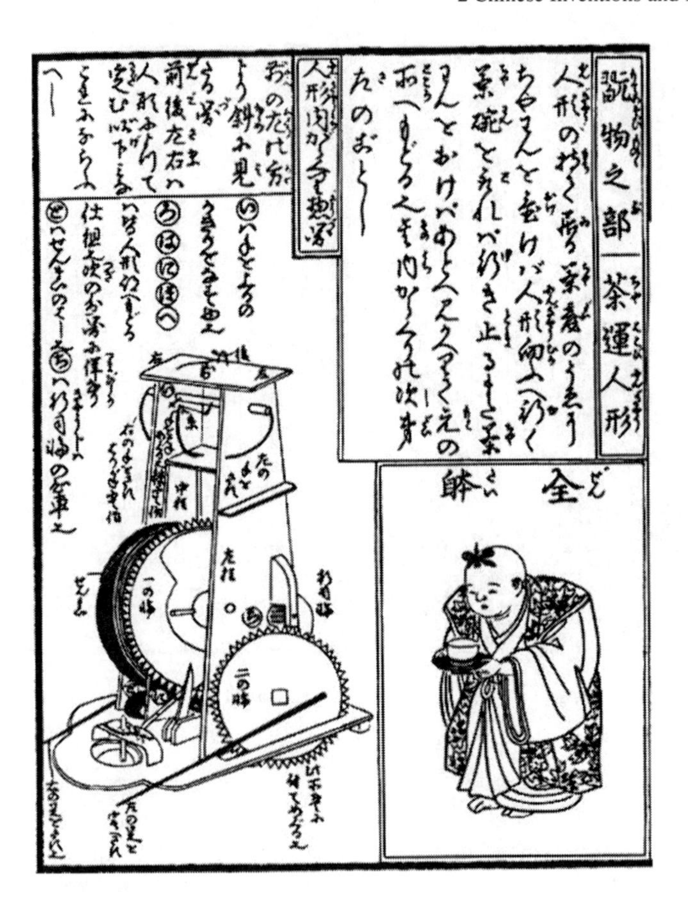

Fig. 2.25 The tea server

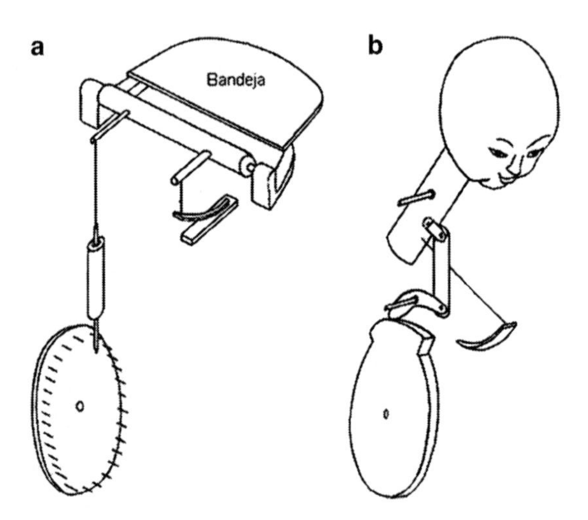

Fig. 2.26 Basic mechanisms in the tea servant automaton in Fig. 2.25: (**a**) Movement starting mechanism, (**b**) Head bowing mechanism

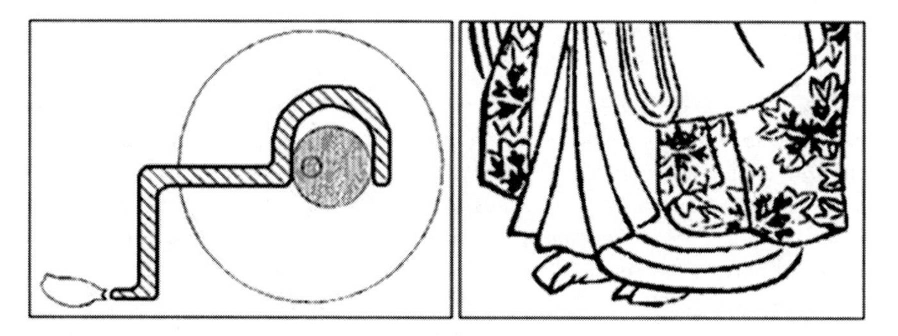

Fig. 2.27 Forward motion mechanism

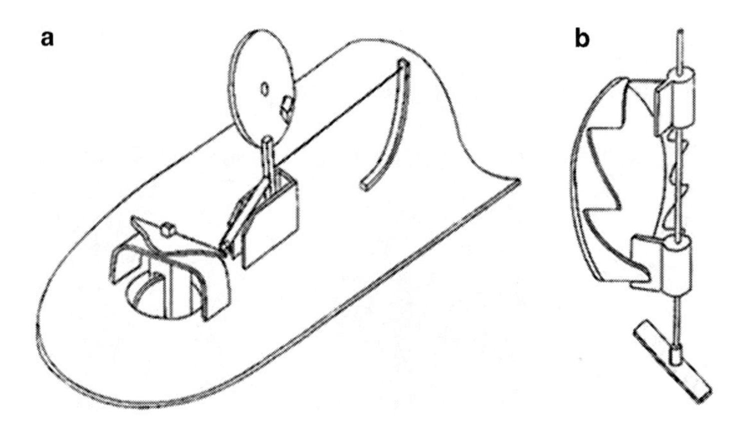

Fig. 2.28 Mechanisms in the tea servant automaton in Fig. 2.25: (a) Turning mechanism; (b) Speed control mechanism

tensed thread (Fig. 2.28a). The lever is composed of an upright rod and another horizontal one that is perpendicular to it whose aim is to make contact with the wheel that determines the movement. When the spin of the driving wheel actuates the projecting part into contact with the lever, the lever will move to push the horizontal rod to exert pressure on the turning platform and it generates the rotary movement of the automaton. The projecting part of the drive wheel needs to be of the exact height and width in order to obtain a 180° movement of the automaton. Once this rotation completed, the pressure that is generated by the wheel and tensed thread, switches the automaton operation to the forward mechanism.

The final mechanism in the automaton controls its speed. This is achieved by a toothed ring wheel which turns intermittently, thanks to two flanges which are shown in the right of Fig. 2.28b. It is a mechanism which is similar to the escapement of a clock.

On Continuity over the Millennia

The previous pages have shown the development of Chinese machinery up to the end of the seventeenth century. Let's compare the designs of the previous automatons with the reconstruction in Fig. 2.29 that is related to a fifth century BC chariot. The mechanism that operates the legs to simulate the horse's gait is remarkable, even more if as the chariot transported heavy loads Marco Ceccarelli wrote in "An Historical Perspective of Robotics Toward the Future" (2001), [33].

It is remarkable how each technical field evolved. Agricultural, hydraulic, military, astronomical, or purely mechanical techniques evolved at a speed that was stimulated by illustrated books as previously described by the examples in the pages of this book, so revealing a technology that was barely accessible beyond its borders.

It is evident that Chinese technical know-how surpassed the engineering skills in Europe or the Islamic world during the same period of time. It is also curious to reflect that some of their discoveries did not reach us (or were not reinvented in the West) until the middle or the end of the eighteenth century, being Chinese Society not involved with those inventions or reinventions.

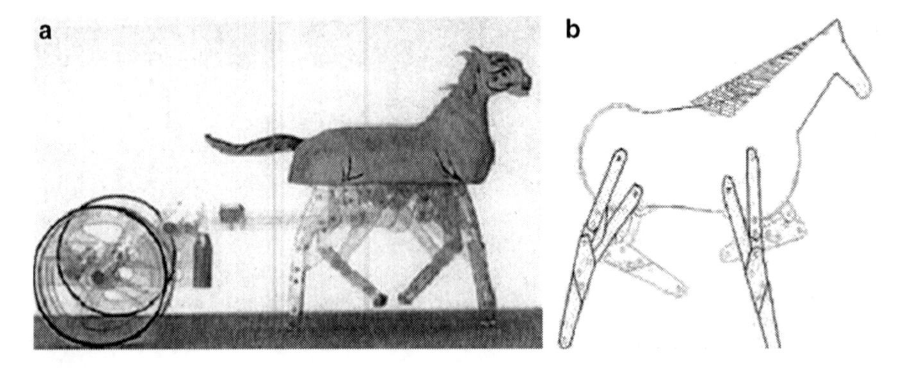

Fig. 2.29 A Chinese automaton chariot of the fifth century BC: (**a**) a diagram of the mechanism; (**b**) Reconstruction of the leg mechanism [129]

Chapter 3
Mechanical Engineering in Antiquity

Like since the first man appeared on the face of the Earth, in Antiquity help was sought to overcome human physical limitations and to make the hard and most difficult tasks easier. In Antiquity, slaves were forced to carry out the most arduous tasks, but solutions were also contemplated in the form of machines or simple devices that would have replaced slaves or would have assisted them in several kinds of work. This machine design practice and activity established the first technical culture of independent competition and professional fields.

The fact is that the majority of automatic mechanisms of Antiquity have not been preserved, but their legacy can be found in some documents, artistic representations, and even in the literature. In general, machine engineering was relevant in Antiquity, mainly in Greece and during the Roman Empire.

The Greco-Roman world absorbed a large part of the technology that was developed by previous cultures, particularly in the Middle East. The Greeks' feelings for philosophy and science were nurtured by the ideas from Asia Minor, the East, and Egypt, and they were expressed in the form of the Hellenistic culture. Roman common, practical sense merged all these cultural backgrounds towards technological use.

Greece reached a high level in the field of technology. An emblematic example was the School of Alexandria in Egypt. From the third century BC in Alexandria there had been intense teaching and research activities on automatic devices. Since the beginning, there were outstanding personalities like Archimedes, Euclid, and Ctesibius, and later in the first century, Hero and Philon worked and taught there.

Hero was a brilliant example of personality in the School of Alexandria, who looked in depth at the different applications of automatons in his treatise entitled "Pneumatics". This treatise became a fundamental reference for automatic machine design in Antiquity and, even later when it was rediscovered, it has been used as an important source for machine design since the Renaissance time.

In fact, several historians consider Hero as the first engineer, since he was the first to produce detailed diagrams in accurate machine drawings. Up to that time, drawings had been more general and in inexact forms, which facilitated their disappearance and oblivion. On the contrary, the high definition of Hero's diagrams can be considered as proof that they are illustrations of machines that really existed.

E.B. Paz et al., *A Brief Illustrated History of Machines and Mechanisms*,
History of Mechanism and Machine Science 10, DOI 10.1007/978-90-481-2512-8_3,

Greek culture evolved and progressed in combination with later Roman technology. The Romans developed a deep technical culture that spread throughout many fields such as civil engineering (roads, bridges, buildings, etc) and military applications (war machines, defence structures, etc.) when necessary, Roman engineers improved mechanical design and automation operation of existing machinery. A brilliant example of those engineers is Vitruvius, who lived in the first century BC and wrote the encyclopaedic treatise "De Architectura", which was rediscovered and its chapter on machines was used as a kind of handbook from the Renaissance on. A later engineer personality is Frontinus who published "De aquaeductu urbis Romae" in the first century after Christ.

On Technological Evidence

Even without any need to refer to written texts, archaeology often provides evidence of machines illustrating the technological know-how of the peoples who built them. Paintings and bas-reliefs indirectly describe the machines that were used in Antiquity, even if the accompanying texts make no reference to technical contents.

As proof of the technological progress of the Egyptian civilisation, the use of lubricants was significant for lowering the resistance to sliding of large blocks of stone and statues. Figure 3.1 shows how a statue was moved during the twelfth Egyptian dynasty. Just in front of the statue there is a slave or operator whose job is to provide lubricant on the surface of the sliding path.

The Egyptians also made widespread use of war machines. One example is the war chariot in Fig. 3.2, whose efficiency depended on an appropriate mechanical design, with particular emphasis on rolling joints.

During the first dynasties of the Egyptian Empire, there is evidence that they had tools for drilling operations that were also aided by an abrasive that impregnated the rocks. This can be noted in Fig. 3.3 showing part of a bas-relief in the Egyptian Museum in Cairo.

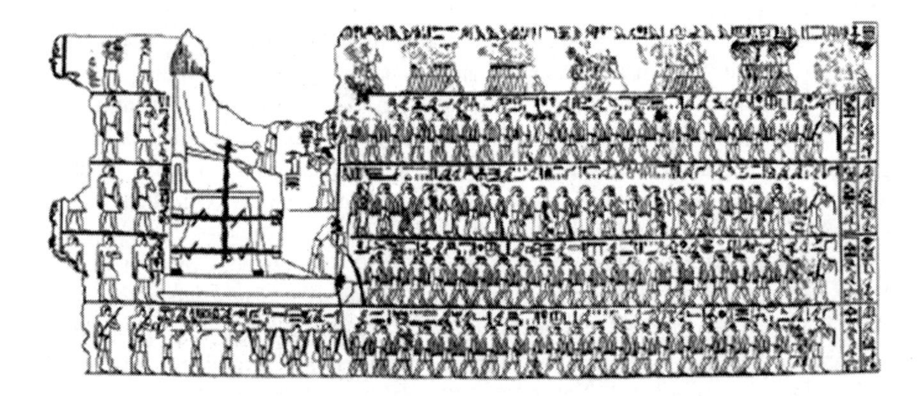

Fig. 3.1 Using a lubricant for moving a statue during the twelfth Egyptian dynasty

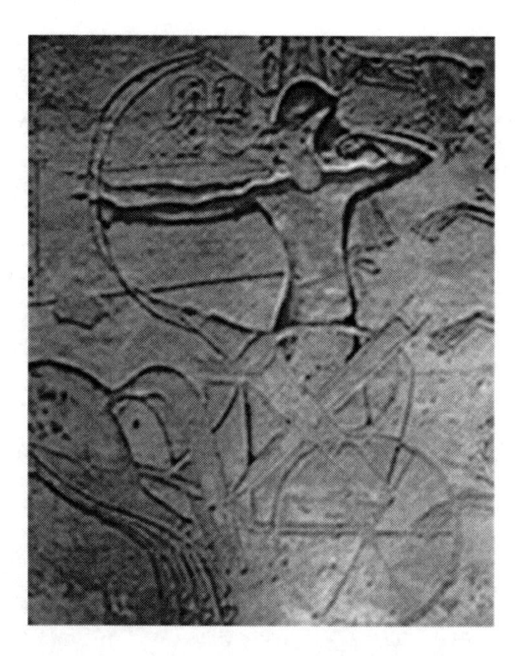

Fig. 3.2 Egyptian war chariot, from a relief of Ramesses II fighting at the battle of Kadesh, at Abu Simbel

There is interesting evidence about the level of knowledge in Antiquity, such as the illustration showing the use of precision scales in Fig. 3.4a, which is part of a relief from the year 1250 BC. Another example is the jar in Fig. 3.4b, showing an interesting drawing of a loom.

On the Development of Ingenious Mechanisms

One of the most relevant incentives for developing mechanisms in ancient cultures was the need for measuring time for several reasons, both with secular and religious aims. Another purpose was to make devices for games and recreation for a sector of society that had time for leisure.

Most of the mechanisms examined are mere curiosities without any apparent usefulness. The ownership of those mechanisms was used to show social prestige so that they even became artistic objects that could additionally produce surprising movements or melodic sounds.

The craftsmen enhanced their ingenuity to supply the market with these luxury goods for top society. Sometimes, the experience that was accumulated from making these automatons was later used to design and build machines for practical uses. The automatons can be considered to be initial experimental prototypes of devices for practical purposes.

Fig. 3.3 Evidence of the use of drilling tools, from a relief in the Egyptian Museum, Cairo

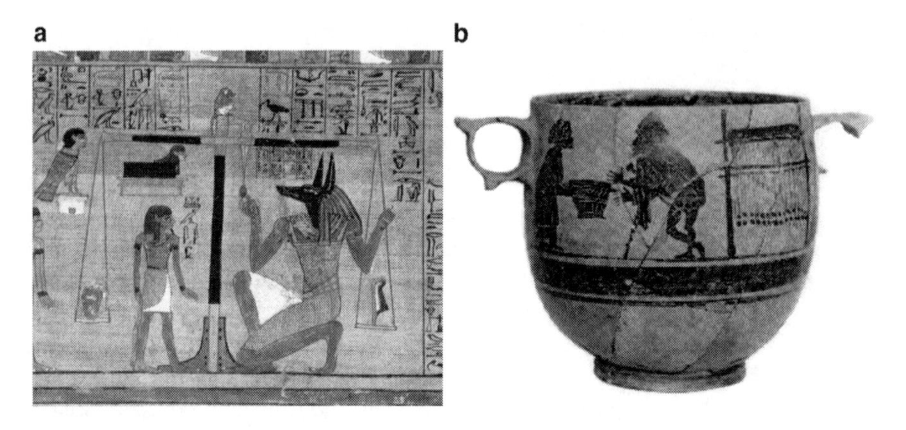

Fig. 3.4 (**a**) Scales in an Egyptian relief, 1250 BC. (**b**) Jar showing a loom, 430 BC [33]

There is proof of the existence of very simple Egyptian water clocks, like the clepsydra in Fig. 3.5, which measured time by using the flow of water from a tiny hole with a constant cross-section. According to the reconstruction by the Jesuit

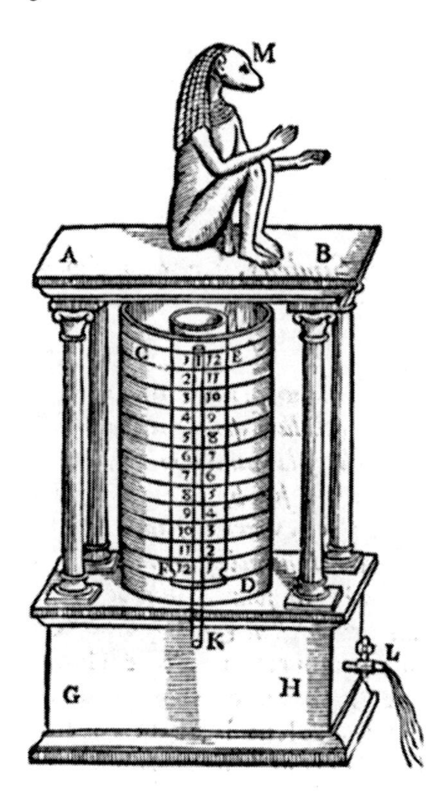

Fig. 3.5 A reconstruction of an Egyptian water clock (Form the work, Oedipus Aegyptiacus, 1652–1654)

priest Athanasius Kircher, in his work Oedipus Aegyptiacus in 1652–1654, it had a graduated scale of hours and a dog-faced humanoid on the top as an ornament.

Giving continuity to the technological progress, the Greco-Roman civilisation continued to develop time-measuring instruments by using water as the power source.

During the third or second century BC, several devices were invented that are attributed to Ctesibius, like the water clock or clepsydra in Fig. 3.6a, according to Vitruvius's opinion that was also reported in the book by Abraham Rees "Clocks, Watches and Chronometers", in 1819. The Ctesibius clock was powered by water falling from a full tank through a pipe to an open cylinder. The cylinder had a floating piston with a rack that moved a pinion with a hand-shaped indicator that turned and pointed to the hour signs.

Ctesibius's main improvement over previous water clocks was his adaptation to Egyptian hours, which were of a different duration according to whether it was day or night. This adaptation was achieved by a cone-shaped device to limit the flow to the cylinder, together with a pipe for discarding excess water. The disadvantage of the system was that the clock needed two manual adjustments every day, namely one in the morning and one in the evening.

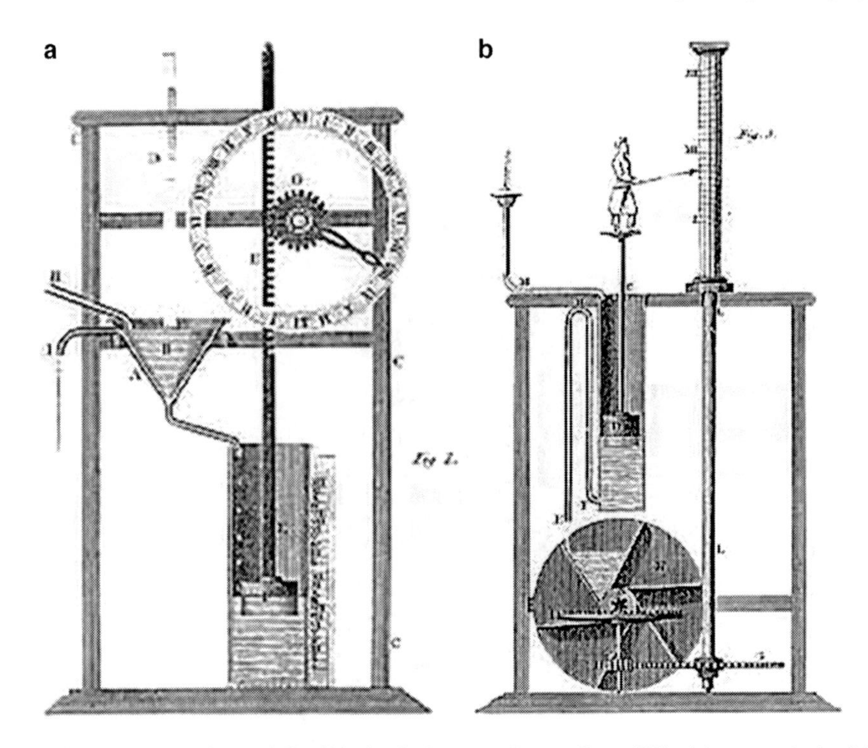

Fig. 3.6 Reconstructions of Ctesibius's clocks according to Rees [96]: (**a**) water clock. (**b**) Improved version

Figure 3.6b shows an improvement of the previous clock that was again attributed to Ctesibius. The clock is decorated with a human figure pointing to the hours on a column. The level of water in the cylinder moves a float up or down, together with the human figure. From the cylinder, the water passed through a U-shaped pipe and fell into a drum that was divided into compartments. When a compartment became full, the drum rotated slowly together with the column marking the hours. This rotation was produced by a gear transmission using several gears with several speed reduction ratios.

Ctesibius used to decorate his machines with moving figures, like automatons. This was also typical of other personalities from the School of Alexandria, such as Hero.

Another very popular machine of the time of the School of Alexandria was Ctesibius's organ that is shown in Fig. 3.7a. According to Dr. Richard Pettigrew, in 1992 Greek archaeologists recovered the fragments and reconstructed an organ that can be dated from the first century B.C.

According to the reconstruction in Fig. 3.7b, it consisted of a series of pipes that were installed on a platform under which there was a pipe for compressed air. The air was compressed by a pedal pump. In order to keep the air pressure constant,

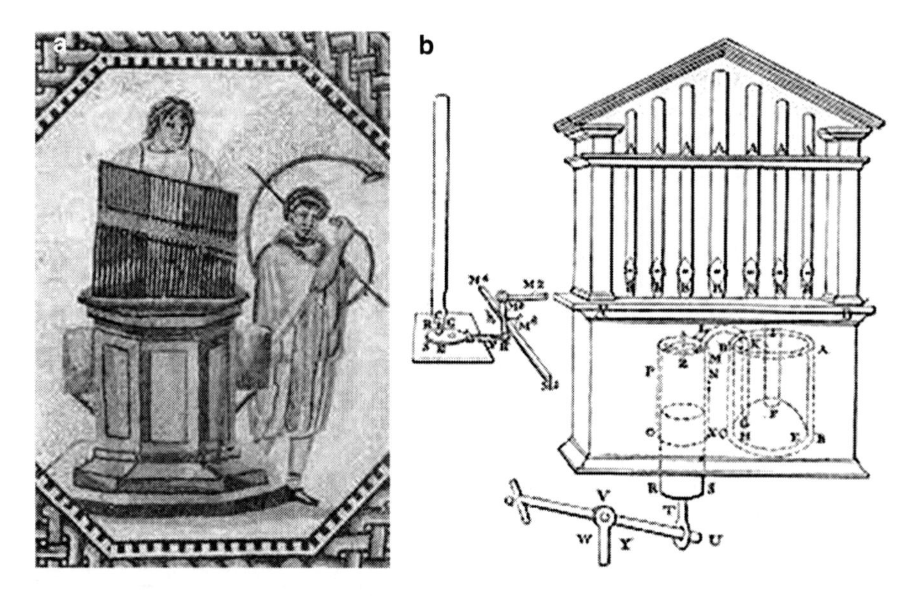

Fig. 3.7 The organ by Ctesibius: (**a**) in an ancient mosaic; (**b**) a reconstruction showing how the keys work (From B. Woodcroft's translation of "Pneumatics" [137])

a tank was submerged in a water container. When air was pumped into the tank, the water passed to an outer container.

Under each pipe was a perforated wooden board. When one of the keys was pressed, an articulated mechanism was lined up with the hole by giving the compressed air in the pipe the possibility to produce a note, as can be noted in Fig. 3.7b. This diagram is taken from the translation of Hero's "Pneumatics" by Woodcroft in 1851. In addition, Woodcroft's work shows continuing technical and historical interest in the machines of Antiquity, which began in the Renaissance and persists to the present yet.

In another version of this organ, the pump did not require manpower, but the operating mechanism is similar to that in a windmill. In this alternative version that is shown in Fig. 3.8, the wind moves a sail, the shaft of which has a wheel with four spokes projecting from it. This shaft moves the vertical piston that pushes the air into the compression chamber. The piston's own weight compresses the air that has entered during the intake stage.

It is remarkable to note the change in the drive energy. The invention of the windmill has been attributed to the Arabs, although they were only really responsible for its spread to the West from its Persian origins, according to the testimony of Hero, who describes windmills already used in that country for milling and water pumping in the seventh century BC. Their use gradually extended to Europe and North Africa via the Arabs and the Crusades. The first reference to the existence of windmills in Europe is due to Ibn Abd el Munim, who made reference to a windmill that was installed in Tarragona in the twelfth century.

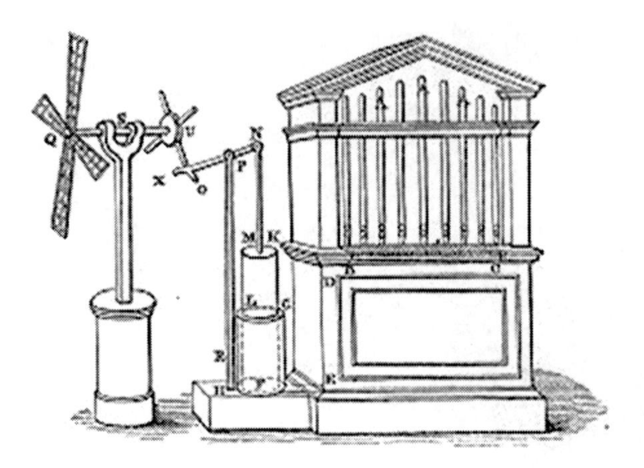

Fig. 3.8 The air-powered organ of Ctesibius (From B. Woodcroft's translation of "Pneumatics" [137])

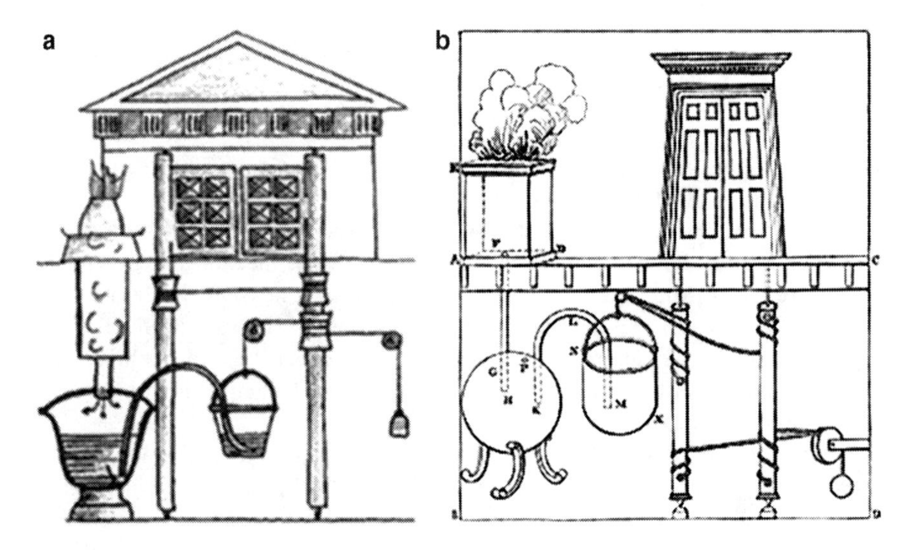

Fig. 3.9 Reconstructions of the automatic opening and closing of doors in a temple: (**a**) colour sketch; (**b**) (From Bennet Woodcroft's translation of "Pneumatics" [137])

Evolving the organ in Fig. 3.8 from a recreational automaton to an industrial use of wind energy was obviously only a matter of scale which, without any doubt, occurred in the West during the Greco-Roman civilisation.

The altar in Fig. 3.9 can be ascribed to Hero, who was a successor of Ctesibius. The altar is described in his work "Pneumática", as an application for automatic doors in the holy precinct of a temple that can be opened and closed automatically. It was provided with a torch to make an offering of fire to the gods. The fire heated the air in a container that was partially filled with water. As the air expanded, the pressure in the container increased by forcing the water to pass through a siphon to a second container.

The container was hung from a pulley that let it fall down to pull a rope operating the rollers that automatically opened the doors. When the fire was extinguished, the process was reversed and the doors automatically closed again. This mechanism is recognised as a significant forerunner of the steam engine. It represents an innovation for the source of drive energy. Its size lets it be called a machine and, more exactly, a heat machine, although its purely liturgical use relegates it to be classified as an automaton or mechanical curiosity.

Without an abundance of slave labour, the Greco-Roman civilisation would probably have developed heat machines at the beginning of our time.

Figure 3.10a shows a reconstruction of Hero's mechanism known as the "divine box" by Bennet Woodcroft. An ingenious mechanism makes the bird at the top turn and to sing by manually turning a wheel. The bird is connected to an upright shaft with a toothed wheel. Movement is transmitted to this shaft by a wheel that is located on the same shaft as the drive wheel. At the same time, this shaft also has a pulley with a bell hanging from it with a chimney. When the bell is submerged into the water, the air flows up to the chimney by making a whistle blow.

Further evidence of Hero's creativity is the "singing birds" machine that is based on the same principle as the "divine box". In Bennet Woodcroft's 1851 reconstruction in Fig. 3.10b, it can be seen that the water spouts out from an angel's mouth and falls through a hole into a large covered container, forcing the air out of it through some pipes with whistles at their ends. When the water in the tank reaches a certain level, it overflows to a second container via a siphon where a floating link

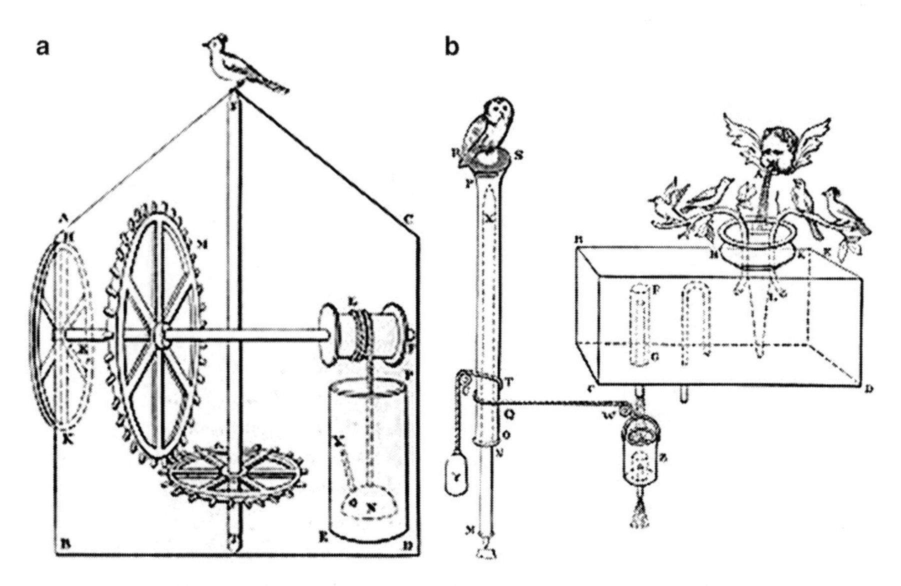

Fig. 3.10 Automatic machines by Hero (From Bennet Woodcroft's translation of "Pneumatics" [137]): (**a**) Hero's "Divine box". (**b**) Hero's "Singing birds"

attached to a rope and counterweight, turns a column with an owl on the top. There is an overflow pipe in this tank that let the owl return to its original position.

On Gears and Screws

The previous illustrations show practically all the elementary mechanisms still used today, namely shafts, couplings, joints, cams, gearwheels, flexible transmissions, hydraulics, and pneumatics.

These were brilliant machine parts for their time, since they were very simply made through rudimentary manufacturing processes. Figure 3.8 is an example where a four-spoke wheel acts as a cam. Figure 3.10a shows another example where the teeth can be understood as primitive teeth of gears.

The amazing Greek developments in theoretical geometry were always applied to both practical and theoretical mechanics. To a large extent, geometry has been also considered as a science of motion and therefore it was applicable for building mechanisms and machines.

Spiral motion was known in Ancient Greece since the time of Archimedes (287–212 BC), who designed the spiral screw for raising water via the gaps between the screw and the outer casing. These machines were widely used and illustrated by the Romans. Renaissance texts include one of the first illustrations of an Archimedes screw, like that in Fig. 3.11a by Honrad Kyeser from his work "Bellifortis", at the beginning of the fifteenth century. Figure 3.11b shows another Renaissance reproduction by Daniele Barbaro.

The construction of gears and particularly interconnecting sections bears a close relation to geometry.

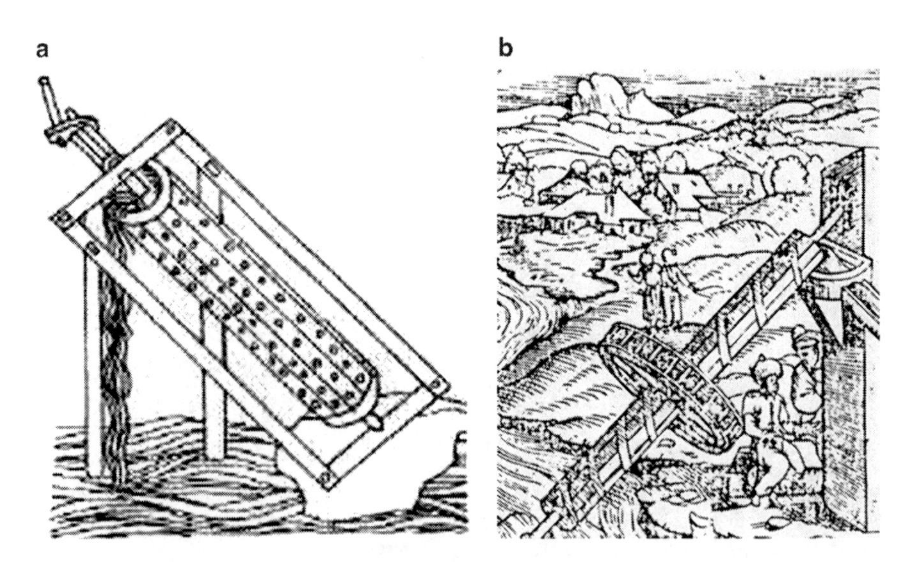

Fig. 3.11 Archimedes' screw pump: (**a**) Kyeser's illustration; (**b**) a drawing by Daniele Barbaro [15]

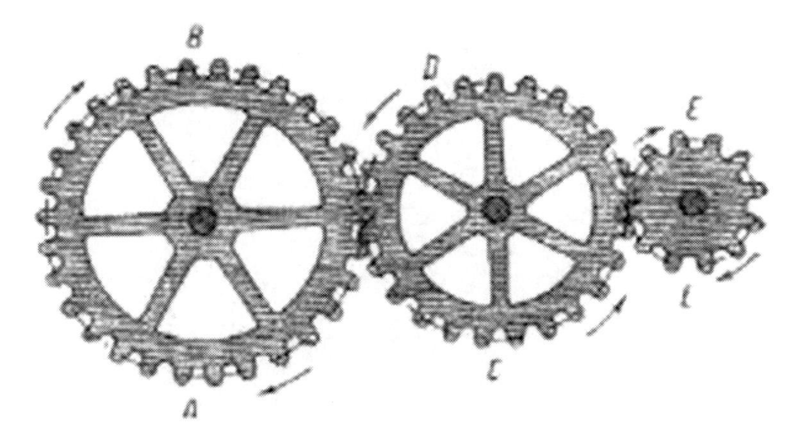

Fig. 3.12 Gear wheels described by Aristotle as in Milonov Ju.K's reconstruction [129]

In the fourth century BC, Aristotle described the transmission of movements by using iron or metal gearwheels. Figure 3.12 shows a reconstruction by Milonov Ju.K. in 1936.

An example of Greek expertise in constructing gear assemblies is the astronomical instrument in Fig. 3.13. It is a front view (Fig. 3.13a) of a mechanical gear assembly from the first century BC, found in the wreck of a Greek vessel in Antikythera (Greece). The reconstruction (Fig. 3.13b) was made by M.T. Wright in 2005.

Combining a gear wheel and a screw led to the worm gear, which required considerable geometric and technical problems to be overcome if it was to work properly.

According to Sigvard Strandh's reconstruction in Fig. 3.14a, Hero's odometer is an example of how such a complex mechanism was used. Figure 3.14b corresponds to one of Leonardo da Vinci's Renaissance machines designed for the same purpose.

It is an instrument for measuring the distance travelled by a vehicle. Movement is transmitted by the lower part via a pinion and has numerous stages of worm-type gear speed-reduction until it reaches the last upright shaft. This shaft moves very slowly and has a container full of balls at top.

The ball container has a hole that lets the balls drop down a motionless vertical tube every time the hole coincides with the hole in the tube. Thus, the speed of the vehicle can be related to the number of balls dropping to the bottom. It is to be noted that, unlike Leonardo's odometer, Hero's allowed greater distances be dealt with and precisely measured due to the use of a worm gear.

On the Way to Mechanical Engineering

Throughout several centuries, mechanical knowledge became systematically applied to devices that supplied considerable force and consumed considerable power. Those devices were real machines.

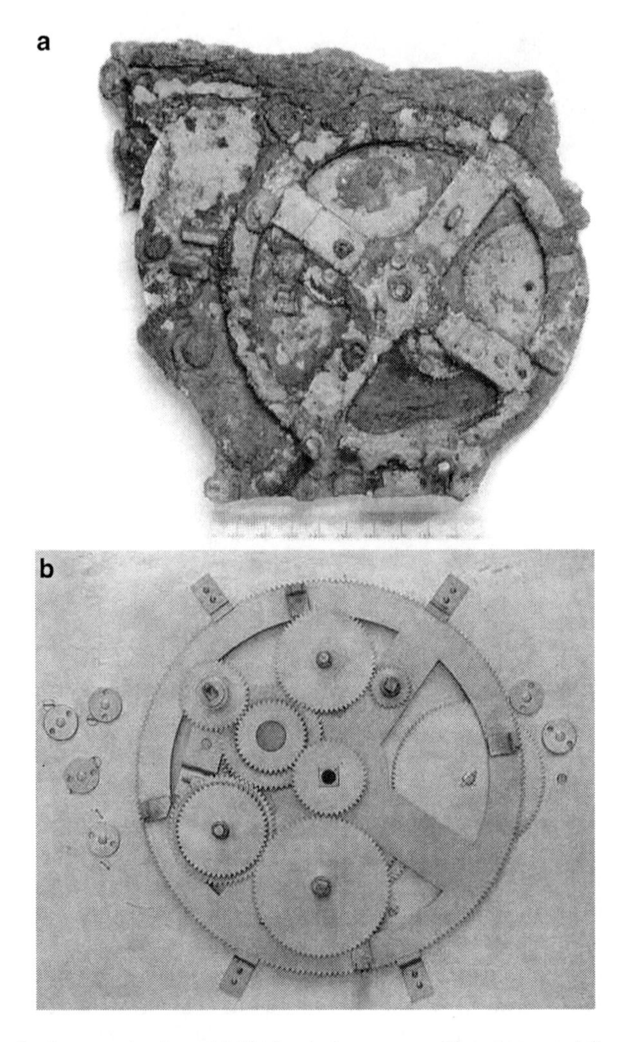

Fig. 3.13 Antikythera mechanism: (**a**) Mechanical gear assembly constructed from the remains, front view. (**b**) M.T. Wright's reconstruction

The activities of military forces during these centuries also stimulated the development of more efficient war machines. Even entertaining events that were attended by large crowds of citizens needed machines whose dimensions required the support of mechanical designs. Mining activity reached proportions that made manual labour unfeasible and therefore it was necessary to pump water from the galleries in amounts that would be described today as on an industrial scale.

All the civilisations coming together under the Greco-Roman world undertook large-scale public works requiring appropriate machinery, even though there was an enormous amount of slave labour. Mechanical engineering gradually became consolidated as a profession that designed, built, and operated machinery.

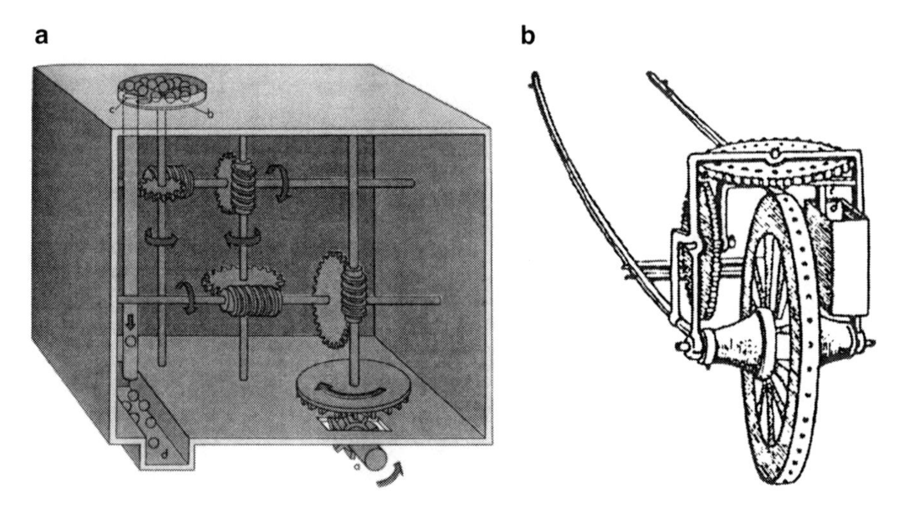

Fig. 3.14 Odometer machine: (**a**) Sigvard Strandh's reproduction of Hero's odometer. (**b**) Leonardo da Vinci's odometer drawn during the Renaissance [119]

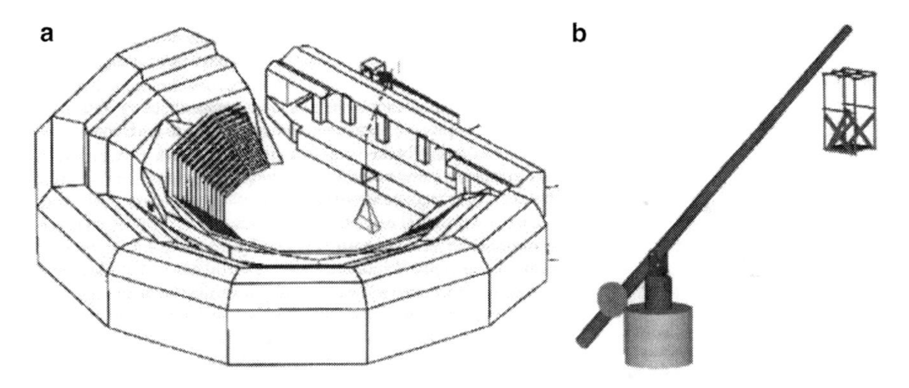

Fig. 3.15 The Mecano mechanism in Greek theatre according to Chondros in 2004 [129]: (**a**) Reconstruction of a fifth century BC Greek theatre. (**b**) Reconstruction of a Mecano

Greek theatre reached its splendour in the fifth century BC. Evidence can be found in classical texts of this period on the existence of mechanisms that were used in theatre, although we have no illustrations. T.G. Chondros's 2004 reconstruction shows an elevator of the period called a Mecano, which is thought to have been able to lift over 500 kg. Figure 3.15 shows a three-dimensional drawing of the ancient theatre in Athens with a Mecano and a description of its parts.

Elevators like those in Fig. 3.16 were also employed in Ancient Egypt, and operated by several people using a rope and pulley. Milonov Ju.K's 1936 reconstruction can be appreciated in this figure.

Fig. 3.16 Elevator used in Ancient Egypt, reconstruction by Milonov Ju.K. [129]

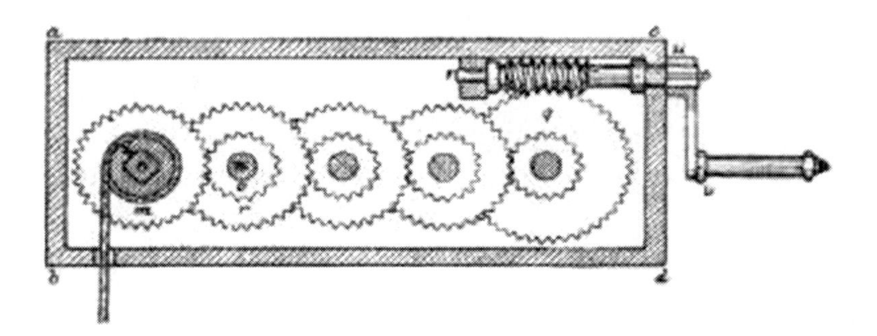

Fig. 3.17 Mechanism designed for raising loads, according to Hero's works. Reconstruction by Milonov Ju.K. [129]

Regarding the same machines, Hero presented a mechanism for lifting loads, whose reconstruction is shown in Fig. 3.17 according to Milonov Ju.K. (1936). This figure shows a worm gear that was designed to obtain a considerable reduction in speed of operation and to multiply the force in lifting operations.

Also relevant are hydraulic machines like Ctesibius's famous suction pump that was described in detail by Hero. The discovery of this kind of pump among Roman remains in Huelva is proof that it was extensively used. It is shown in Fig. 3.18a.

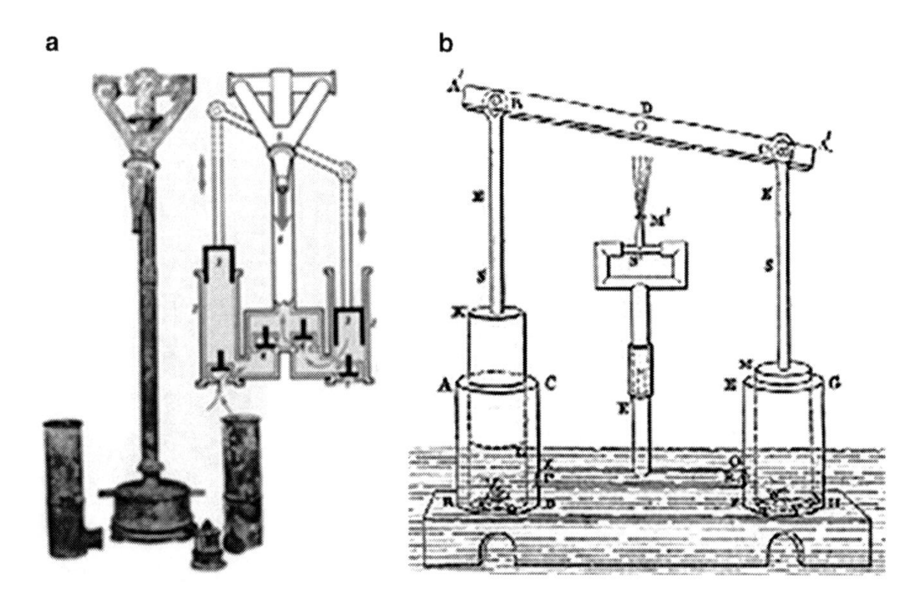

Fig. 3.18 Mechanism design in pump machines: (**a**) Remains of a Roman pump from the Valverde mine in Huelva, Spain. (**b**) Reconstruction of Ctesibius's pump as described by Hero from the translation of "Pneumatics" by Bennet Woodcroft [137]

The pump is operated by a lever mechanism that moves the pistons in two cylinders in order to pressurize water for several purposes like drainage, fountains feeding, fires extinction, etc. As can be noted in Fig. 3.18b, the cylinders were interconnected by mean of valves and through a common vertical pipe for the output of the water.

At the bottom of the cylinders there were connecting valves with a tank. The lever mechanism operated in such a way that when it guided a cylinder to rise, the other cylinder descends. The rising cylinder takes in water through the valve at the bottom and the descending cylinder pushes out the water into the upward vertical pipe, while the water pressure closes the other valves.

Figure 3.19 shows two examples of water-powered machines from Antiquity. They were used for several purposes, such as raising water, mills, etc. Figure 3.19a shows a sophisticated water wheel that is attributed to Philon, as in the reconstruction in the book "A History of the Machine" by Sigvard Strandh. It consisted of a water wheel driving a chain that gives a rotation motion to an upper shaft of a triangular drum. The chain is provided with suitable buckets. As the shaft rotates, the buckets are filled with water from the bottom flow and then they are poured out into a pipe at the top.

The machine in Fig. 3.19b is a gear mill that is described by Vitruvius and is shown as in the book "A History of the Machine" by Sigvard Strandh.

Many other machines implemented a knowledge of the screw and examples can be outlined from different kinds of presses that were described by Hero. Sigvard Strandh's reconstructions show examples for crushing fruit. In Fig. 3.20a, the screw

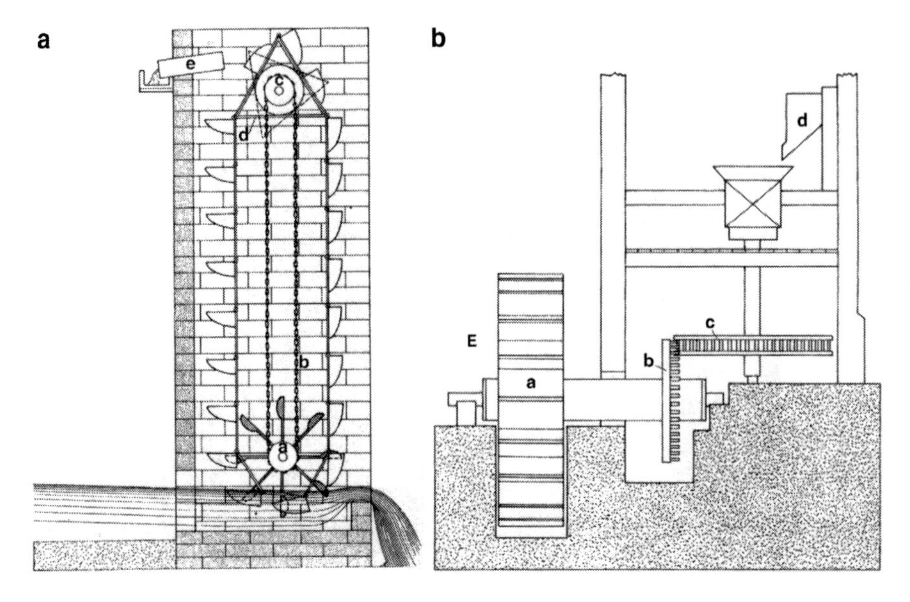

Fig. 3.19 Water-powered machines according to reconstructions by Sigvard Strandh [119]: (**a**) Philon's water wheel. (**b**) Vitruvius' gear mill

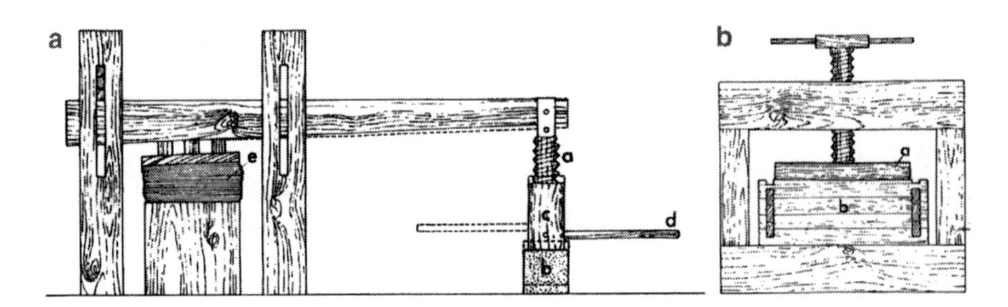

Fig. 3.20 Hero's press; reconstructions by Sigvard Strandh. (**a**) Beam press. (**b**) Direct press [119]

operates a crossbeam for compression by moving the beam vertically. Figure 3.20b shows a direct screw press. This type of press, with some innovations, was found during the Renaissance and during the Iberian Empire.

Among war machines at the beginning of the second century AD, the role of Apollodorus of Damascus is remarkable. His work on siege machines was later reinterpreted in Byzantium. More detailed drawings were outlined with the figures of persons to give some idea of the size of the machines. Figure 3.21 shows a siege tower from a sixteenth century Italian copy of a drawing taken from an eleventh century Greek manuscript by a Byzantine author who, under the pseudonym of Hero of Byzantium, based his work on Apollodorus's manual.

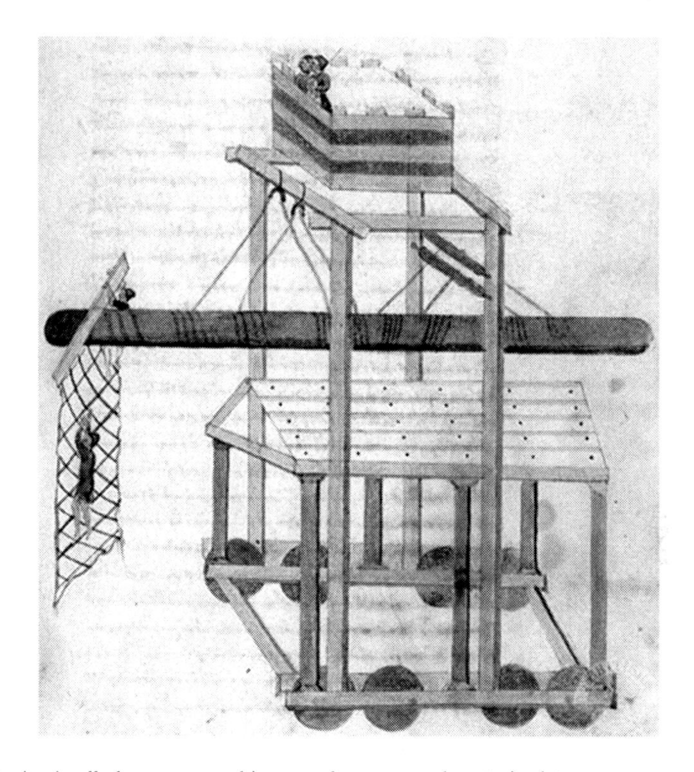

Fig. 3.21 An Apollodorus war machine as redrawn several centuries later

On Vitruvius's Influence

The Roman Empire acquired and improved this development of mechanical engineering during the millennium of the Empire.

When he was young, Marcus Vitruvius worked as an engineer for Julius Caesar, and then he focused his attention on civil architecture in his masterpiece work "De Architectura", which is a compilation of architectural knowledge that also devotes several sections to machines for building activity.

Later, in his work "De aquaeductu urbis Romae", Frontino gathered together the techniques for supplying and distributing water to the capital of the Empire.

In the next centuries, mechanical engineering languished as a professional activity and was relegated to specific spheres, until it was once boosted because of new economic and social conditions. An indicator of the extraordinary level that was reached in the Greco-Roman period was the enormous value that was credited to Vitruvius's work centuries later when, during the Renaissance, society once again began to approach similar technological problems.

There is evidence that the work of Vitruvius was studied by Iocundo in 1511, with a folio edition containing 136 illustrations. There were many translations and

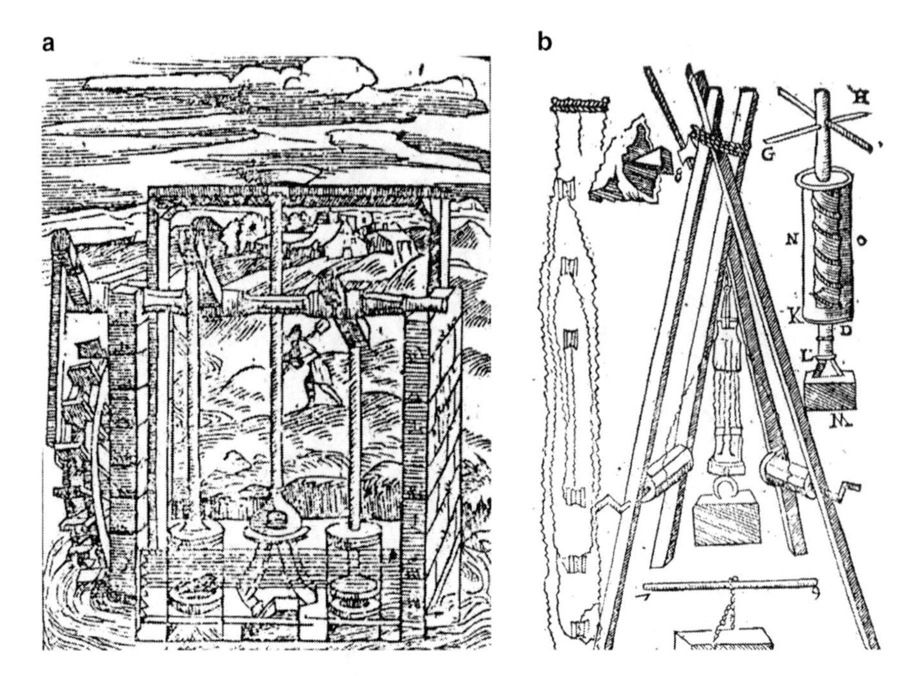

Fig. 3.22 Vitruvius's reconstructions of machines by Danielle Barbaro [15,135]: (**a**) Suction mill; (**b**) Crane

interpretations during the Renaissance and later periods. These Vitruvius editions were published with a wealth of magnificent engravings, like in Danielle Barbaro's work in 1584. Figure 3.22a, b show a water pump and a crane.

Two hundred years later, a machine treatise was published on Vitruvius machines and then it was translated from the Latin and commented on in Madrid in 1787 by Joseph Ortiz y Sanz, who was a priest in the service of the King. This publication includes some more thorough reconstructions of the water raising machines described by Vitruvius, such as tympanums and water wheels or screws by Archimedes, as shown in Fig. 3.23. It can be noted that Ortiz y Sanz also provided some schematic-type drawings for building this screw.

The great works of Roman architecture required the use of several machines and elevators for construction work. These machines were different in the way weights were lifted and in the number of men required to operate them. The machines make use of mechanical parts such as pulleys, winches, ropes, and wheels, with a place for a person to work inside. Outstanding examples are shown in Figs. 3.24 and 3.25.

On Harmony in Machines

The Greco-Roman world had a profound sense of beauty and harmony, which was reflected in its works, even in machines.

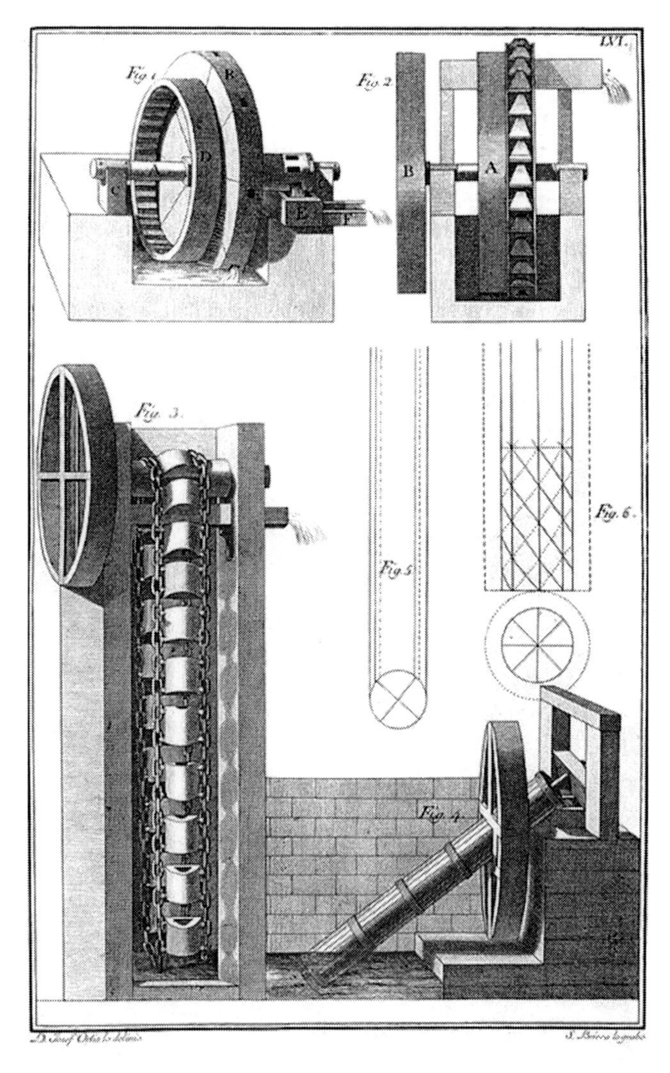

Fig. 3.23 A reconstruction of Vitruvius's machine for raising water in engraving VI by Joseph Ortiz y Sanz [135]

Harmony with the universe, with the nearby surroundings, and between the parts composing a machine. All this was considered, even in the design aspects. But apart from aesthetic considerations, this harmony was aimed at efficient operation of whole. Something in harmony will more likely achieve this purpose.

Geometric criteria of harmony were applied to machines which had to be built in the right proportions. A proportion is made up of ratios and a ratio is a comparison between two sizes, quantities, qualities, or similar concepts, and it is expressed by the formula a/b. Therefore, a ratio consists of the measurement of a difference; a difference to which at least one of our sensorial faculties can respond. The world

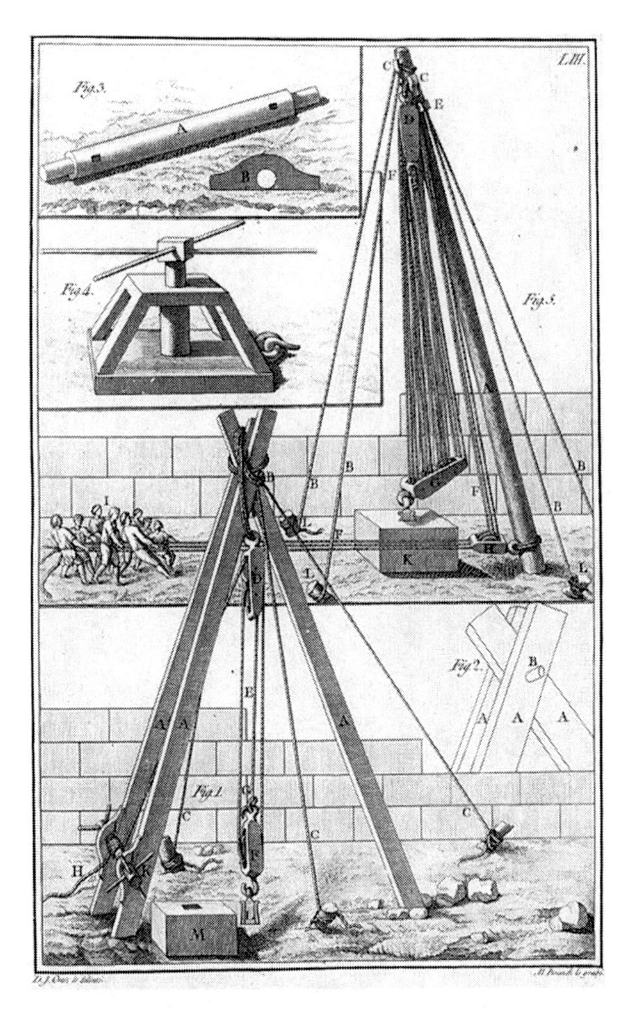

Fig. 3.24 A reconstruction of Vitruvius's machine for lifting loads in engraving III by Joseph Ortiz y Sanz [135]

we perceive is made up of intricate inter-related patterns which Gregory Bateson calls "differences that make a difference".

The golden proportion (usually denoted Φ) is a constant ratio that is derived from a geometric relationship, which like π and other constants, is "irrational" in numerical terms. In one sense, the golden proportion may be deemed supra-rational or transcendental. For this reason, for the Greeks it was a proportion that was the foundation of the experience of knowledge *(logos)*.

It may be said that wherever function is intensified or where there is a special beauty or harmony of form, the golden number will be found.

Fig. 3.25 A reconstruction of Vitruvius's machine for lifting loads in engraving IV by Joseph Ortiz y Sanz [135]

However, it is in the human body where the metaphysical meaning of Φ, can be discovered, as expressed by Heraclites' aphorism: "Man is the measure of all things". According to the different traditions which identify a human model, that is a definition of the average ideal proportions of the body, the body is divided by the navel, according to the golden section (Fig. 3.26).

The parts composing Greco-Roman machines also abided by this concept of proportion. Machines, man, and the firmaments had to be one of the same harmony.

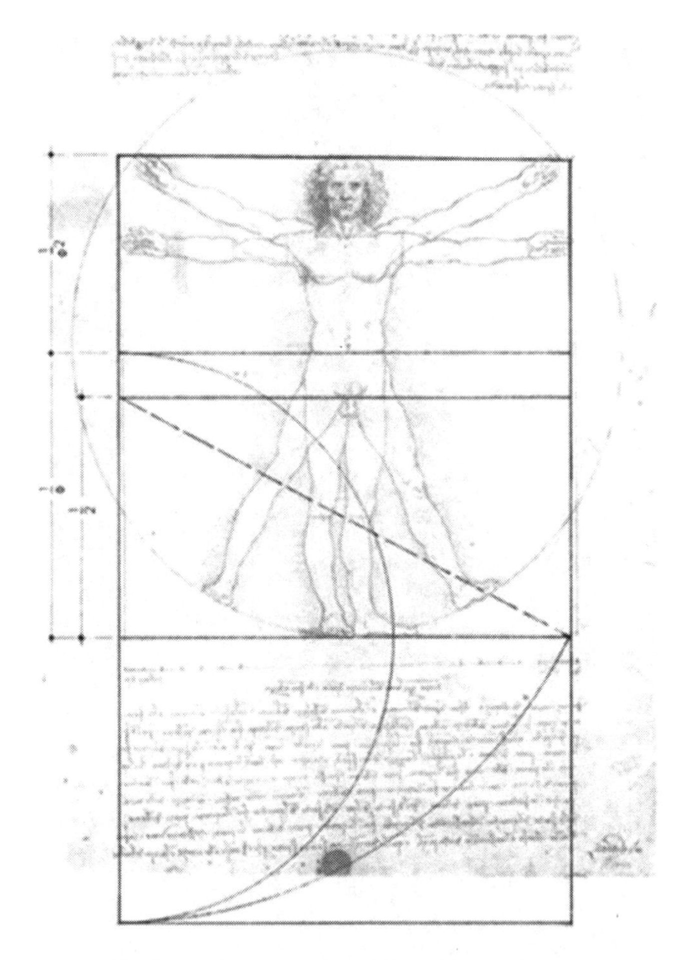

Fig. 3.26 Division of the human body according to the golden section

Chapter 4
Medieval Machines and Mechanisms

In medieval times the most creative mechanical developments took place in the Islamic world. Some of those engineering and technological achievements are little-known due to the fact that, at that time, knowledge passed from master to apprentice through direct experience without being recorded in any written form. In addition, few manuscripts were actually written and only a few of them have survived through time.

The first book of the Islamic world that is known with descriptions of mechanisms was written by the three Banu Musa brothers as a "Treatise on Ingenious Devices". Around the ninth century, together they wrote over 20 works on many several subjects, but their best known work was on mechanisms.

Many of the mechanisms in the book are obtained from a translation of the works of Hero and Philo. In addition, among the 100 mechanisms that are described in the book, there are many that are related to other new machines and even they are more advanced machines than those of Hero and Philo. The Banu Musa brothers focused attention on improving ancient machines like, for example, by adding a series of siphons in Philo's oil lamp to make it more effective. At the same time they designed new machines such as fountain mechanisms, dredgers, scientific tools, agricultural machines, toys, and other automatic mechanisms whose operation is based on valve-based mechanisms, action-reaction systems, and principles of automatic control that demonstrated their creative minds. When those machines are compared to Hero or Philo's works there is less theoretical content but greater engineering ingenuity.

Each description starts with a brief summary of the purpose and motion capability of each device and then a drawing is introduced with an explanation of how it is built. The description is completed by detailed information on how it works. The major drawback is that they almost completely fail to comment on the materials, the manufacturing techniques, or the size of the machine.

The ideas that were developed for the mechanisms in the Banu Musa brothers' book were adopted by later Arab engineers, who took them as a starting point for making the first water clocks. Successively, they transmitted this culture to Spain where the Arabs founded Al-Andalus and from there it spread to the rest of Europe.

E.B. Paz et al., *A Brief Illustrated History of Machines and Mechanisms*,
History of Mechanism and Machine Science 10, DOI 10.1007/978-90-481-2512-8_4,

In 1206, Al-Jazari listed a large number of machines in his book, entitled a "Treatise on the Knowledge of Mechanisms". While the explanations and the drawings are even more detailed than in the Banu Musa brothers' book, as Al-Jazari himself remarks, he seeks wider-ranging development with clearer explanations of the mechanisms and their drawings.

Al-Jazari was the most outstanding mechanical engineer of his time. This is also proved by the subsequent influence he had, and not just in the Islamic world. Many of the machines, mechanisms, or techniques that first appeared in Al-Jazari's treatise later became a part of European mechanical engineering. Some of his design contributions were on double action pumps with suction pipes, the precise calibration of holes, procedures for laminating wood to reduce wearpage, static wheel balancing, and use of paper models for design testing.

Al-Jazari divided his designs into two branches according to their technology, namely "quality" technology and "utilitarian" technology. The "quality" technology referred to machines or instruments that were designed for use by aristocrats or for scientific use. Those machines were made with strict rigour and impeccable care, even for their aesthetic appearance. Clocks and measuring devices were included among these objects. "Utilitarian" technology included machines that served for manufacturing and for workers' economy and were technically much simpler than the "quality" machines, such as mills, water-raising devices or textile machinery.

Although the Banu Musa brothers's and Al-Jazari's books are most outstanding in clocks and automatons, many other books from the Islamic world contain descriptions of former machine collections, like "The Book on the Making of Clocks and their Use" by Al-Saati Al-Khurasani (1203) and "The Book of Secrets about the Results of Thoughts" by Al-Muradi (eleventh century).

While these devices were developed in the Arab world, it was Villard de Honnecourt of France who is considered to be a forerunner of the Renaissance. This architect travelled throughout France in the first half of the thirteenth century and recorded details of building techniques and mechanical devices in his notebooks. At that time, an architect can be understood to be a combination of artist and engineer who conceives and designs a building as a whole with all its details and supervises the construction work, including the machines for it.

In the western world, technology and machines were only important within very select communities and therefore they did not require dissemination or any real technical awareness. Thus, machines and mechanisms were developed by applying past traditions and trying new solutions for specific needs. However, an intellectual approach predominated that gave no importance to technology or machines.

On Raising Water

During the seventh to the fifteenth centuries, the Muslim population was concentrated along river valleys and depended on irrigation for agriculture. The need to use river waters stimulated the invention of several water-raising devices.

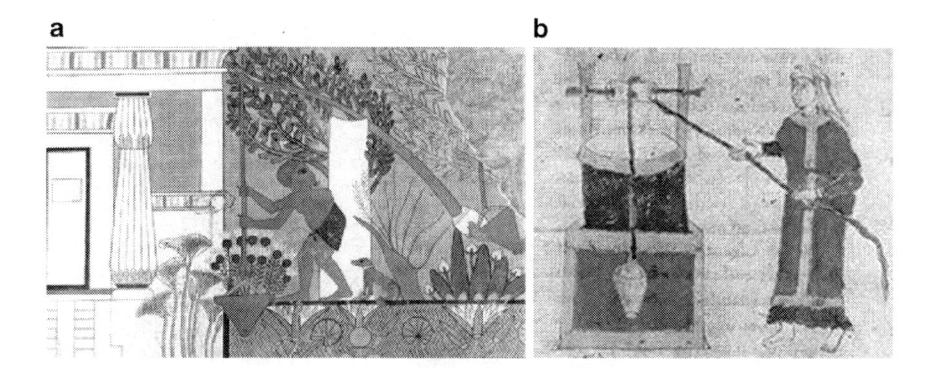

Fig. 4.1 Mechanisms for water raising: (**a**) Illustrations showing the use of the shaduf from the era of Ramses II; (**b**) Illustration from the encyclopaedia of Hrabanus Maurus, "De Rerum Naturis" [78]

The Islamic illustrations of the period make reference to some simple instruments, like shaduf, noria, and the waterwheel. Figure 4.1a is part of a painting from the tomb of Ipuy in Deir-el-Medina and dates from the era of Ramses II, and shows the use of a shaduf as a human-powered tool that was known in China, Egypt, and Syria for more than 4,000 years (Chapters 1 and 2). It consists of a rod with a counterweight that lifts a water container attached to it by an upright pole or rope. The mechanism's simplicity makes it possible to use local materials.

Since it is easier for a man to exert a downward vertical action, the counterweight is used to operate the lifting of the water weight and the human operator pushes the shaduf rod down. Once the water has been raised, it can be used to fill irrigation channels or reservoirs, or even can be channeled to higher ground by using several shadoofs in series.

Another very simple water-raising device can be noted in the drawing in Figure 4.1b, where water raising is achieved by a rope and pulley. The illustration is taken from the encyclopedia "De Rerum Naturis" from circa 1022 by Hrabanus Maurus, preserved in the Abbey of Montecassino. In this case, it can be noted that raising water requires greater human effort since there is no counterweight as in the shaduf.

As regarding the waterwheel and the noria, the main difference can be identified in the source of the driving force, which is often obtained by animals in the noria, while with the waterwheel, it is the force of the water flow that turns the wheel to transport the water. In general, a noria is used for lifting water from wells, while the waterwheel can be used in fast-flowing rivers.

If Fig. 4.2 is compared with Fig. 2.8, the difference can be well appreciated. With the waterwheel, we can see that the wheel, which is powered by water flow, there is no need for manual driven power. However, in the noria, a beam is connected to the wheel shaft that generates the movement supplied by an animal through the required force.

Fig. 4.2 Illustration of a horse-driven waterwheel

The shaduf, the noria, and the water wheel do not represent any significant innovations with respect to the devices pre-existing in China (Chapter 2), although, in the Arab world, they were built more perfectly, maybe due to a greater need for efficiency in the use of water resources.

The next illustrations from the treatise by Al-Jazari provide technical solutions of a greater constructional complexity that reveal a greater capacity than in previous cultures for ingenuity relating to the use of hydraulic machinery.

For some erudite current scholars of this period of history, the author's aesthetic representation and beautiful, careful presentation of some of his designs are due to the fact that those machines are Utopian and they never existed at that time. However, other scholars consider their aesthetic design features to be of a practical nature, derived from the need to improve the efficiency of water-raising methods by conceiving mechanisms for this purpose. Regardless of the foregoing, their designs have the added significance of incorporating important techniques and components for developing machine technology.

Among other machines, Al-Jazari presents a spectacular mechanism in Fig. 4.3 with two continuous operation sub-systems that are operated by animal power. The first sub-system is represented in the lower part of the illustration. A water tank has a hole through which the water falls in a way that causes a bucket-carrying wheel to rotate together with a horizontal shaft. This horizontal shaft is fitted with a gear-wheel that engages another horizontal wheel that transmits the motion to an upright vertical shaft.

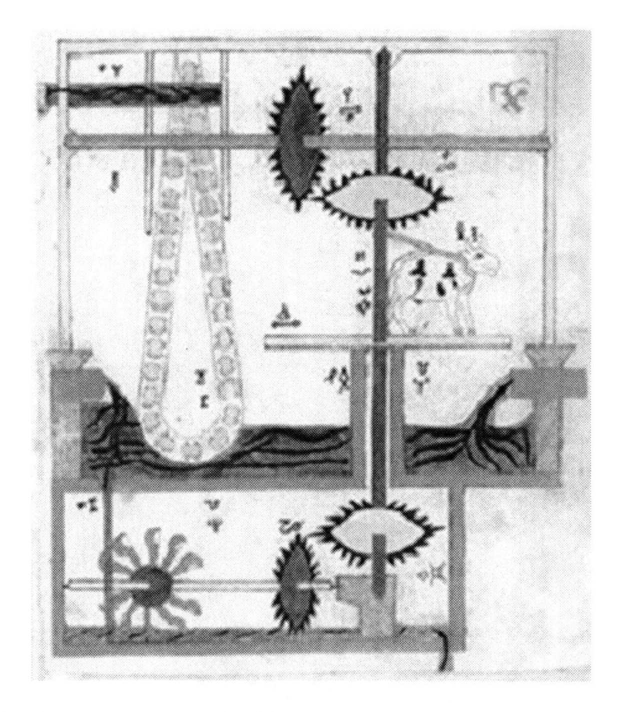

Fig. 4.3 Al-Jazari's water raising device [9]

The second sub-system has two driving forces, namely one is obtained using an animal that is tied to the shaft and a latter one comes from the movement of the bottom sub-system. Both forces rotate the upper vertical gearwheel that operates a second horizontal shaft to raise water from the tank to an aqueduct by means of containers that are connected to ropes rotating around a pulley.

Al-Jazari also shows designs, that were based on tilting movements as an innovation for water raising procedures, several centuries before Juanelo, who also built a machine with tilting buckets to supply Toledo with water.

The machine in Fig. 4.4 shows the use of lantern-type gear wheels. It is composed of two large vertical columns that are placed in a tank to house two cross-shafts. The lower shaft incorporates a scoop and a lantern. The upper shaft has two gear wheels. One is made with teeth on a quarter of its perimeter only, and it engages the lantern, while the second wheel receives the movement from an animal-powered horizontal gearwheel.

When the animal moves, the partially toothed gear is rotated. During the toothed quarter turn, the lower lantern engages the wheel that rotates the lower shaft, forcing the scoop to rise. During the other three-quarter turn, the scoop falls and picks up water so that the cycle starts again when the lantern re-engages the gearwheel.

In the next figures, Al-Jazari obtains a tilting movement by using sliding crank-operated mechanisms. This is probably a more efficient solution than using partially toothed wheels, which would lead to large fluctuations in the required torque.

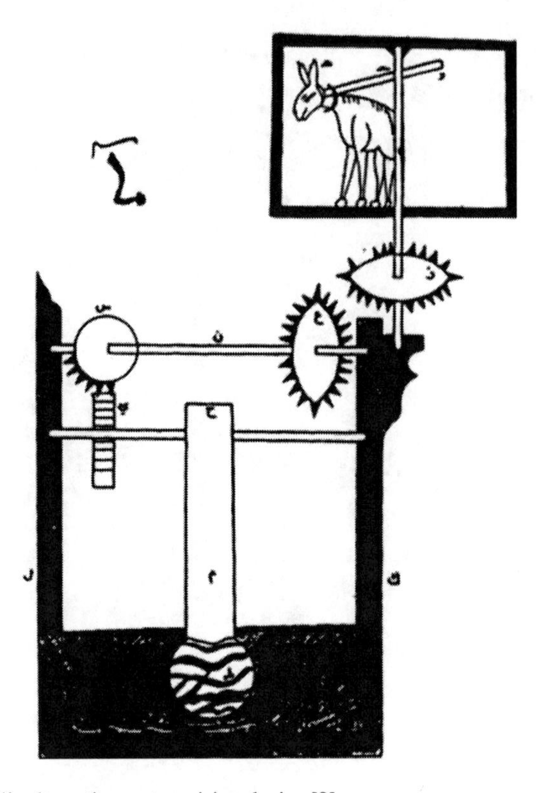

Fig. 4.4 Al-Jazari's alternating water raising device [9]

Figure 4.5 shows another water-raising device by Al-Jazari. A circular movement that is obtained using animal traction is again used to rotate a gear assembly with two wheels on perpendicular shafts. It can be noted that the horizontal shaft incorporates a crank whose free end is housed in a groove made of a scoop in the upper part and the lower part is submerged into water. The turning movement of the crank pushes the scoop to rise and then to fall in a tilting movement raising the water. How the device worked can be better understood by referring to Donald R. Hill's reconstructions in Figs. 4.6 and 4.7.

Figure 4.6 shows the scoop when submerged into water with an ascending stroke. In Fig. 4.7, the spoon is fully raised out of the water and is now ready to empty the water via a tilting channel. In the last movement, the scoop returns to the water to begin the cycle once again.

Al-Jazari mentions other water-raising devices that make use of alternating movement through crank-slider mechanisms.

In Chapter 3, it was discussed how the suction-impulsion pump was already used in the Greco-Roman world. Figure 3.18 would suggest manual operation for this type of water pumping as well as the fact that it may have been used in mining galleries.

Fig. 4.5 Al-Jazari's water raising device [9]

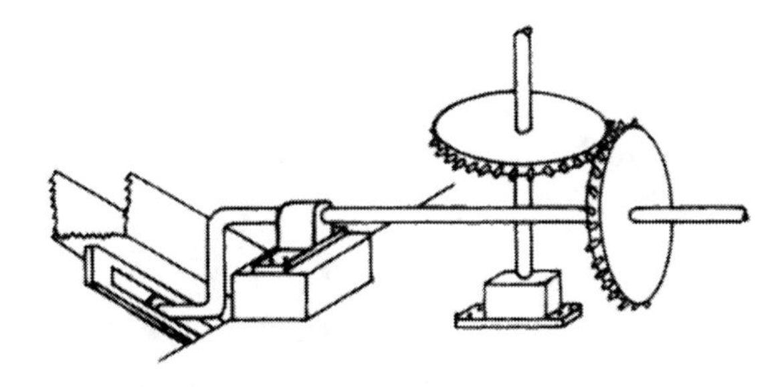

Fig. 4.6 Detail of the device in Fig. 4.5. Donald Hill's reconstruction [9]

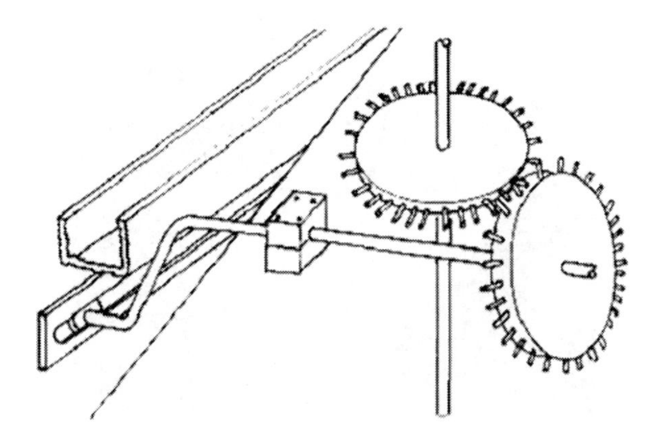

Fig. 4.7 Detail of the device in Fig. 4.5. Donald Hill's reconstruction [9]

The innovation in Al-Jazari's treatise consists in the fact that it is the water flow that is the driving force, as can be observed in the following.

An example of this is the double-effect pump in Fig. 4.8. This machine incorporated mechanisms from two previous machines, namely the paddle-wheel operating the noria and the siphon-pumping device that was described by Philo. It seems a practical device that is aimed at solving a specific problem for raising water cheaply and efficiently from deep rivers up to the surrounding fields and settlements.

Coupled with the gearwheel at the bottom, a lug can be noted that slides like a cross-piece, in the groove of a slide, whose lower end is pivoted at the bottom of the machine. Attached to both sides of the slider are connecting rods that are connected to the head of the two pistons. The copper cylinders contain pipes for suction and outflow. Those pipes connect the two cylinders and the latter operates as non-return valves. The two supply pipes come together as one in the upper part of the machine.

The water-powered paddle wheel operates a gear assembly that transmits movement to a second shaft. The rotation of the gear wheel in the second shaft produces the tilting of the slide by using the lug-slide, This moves the attached pistons from side to side with an alternating movement, so that when one cylinder is sucking, the other is emptying. This motion produces a fairly uniform out-flow through the outlet pipe.

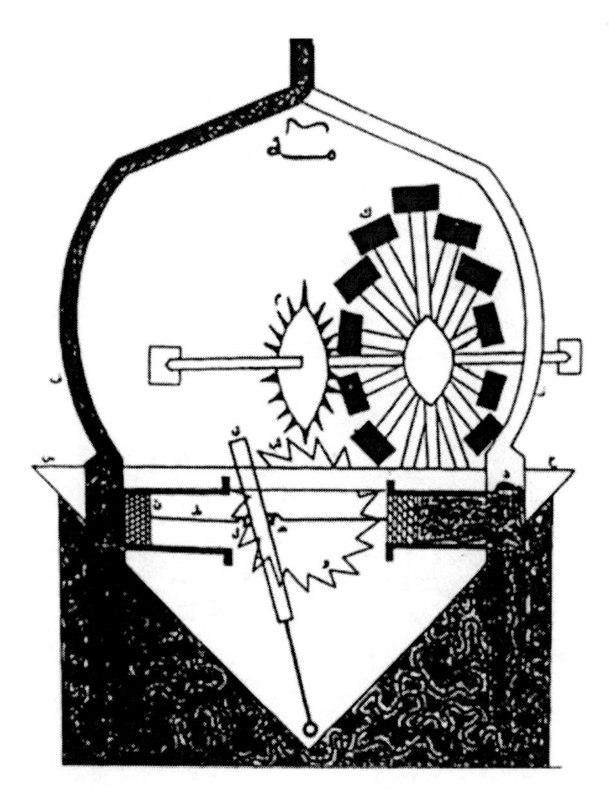

Fig. 4.8 Al-Jazari's pumping device for raising water [9]

Later, in 1551, Taqi al Din, in his book "The Sublime Methods of Spiritual Machines", presented a more productive machine for raising water more uniformly. The machine is based on a common six-piston transmission. It is the so-called six-cylinder mono-block pump that is shown in Fig. 4.9, as an evolution of Al-Jazari's primitive machine.

The water flow drives a transmission with six cams. Each cam governs the movement of one of the six pistons. A suitable angular bias of the cams on the shaft achieves an uncoupled movement of the pistons and a more uniform water flow in the pumping pipe.

When the water flow was insufficient, the alternating movement produced by the action of the water flow stimulated Al-Jazari to other brilliant designs. Among his designs, one of the most noteworthy solutions is the fountain in Fig. 4.10 which forms part of his "utilitarian" technology.

This fountain was fed with water from the top with a pipe that emptied into a tank on either side. The ends of the pipes were open and, near to each end, there

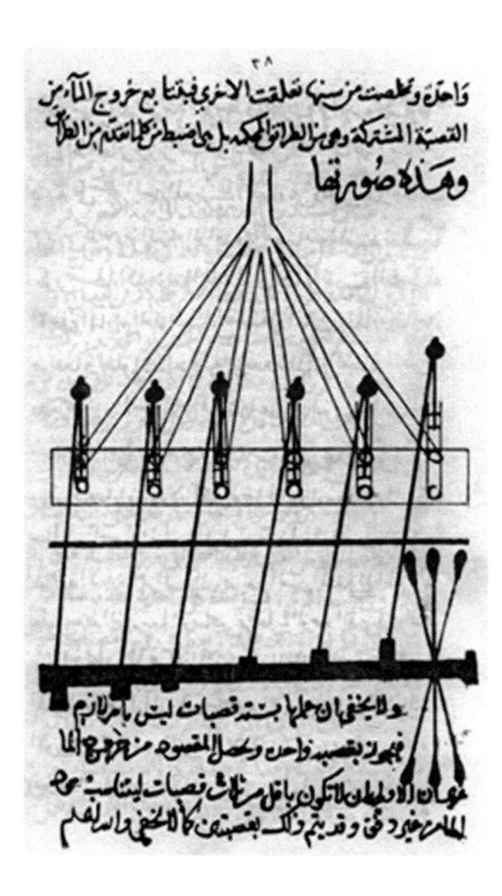

Fig. 4.9 Taqi al Din's water raising machine [7]

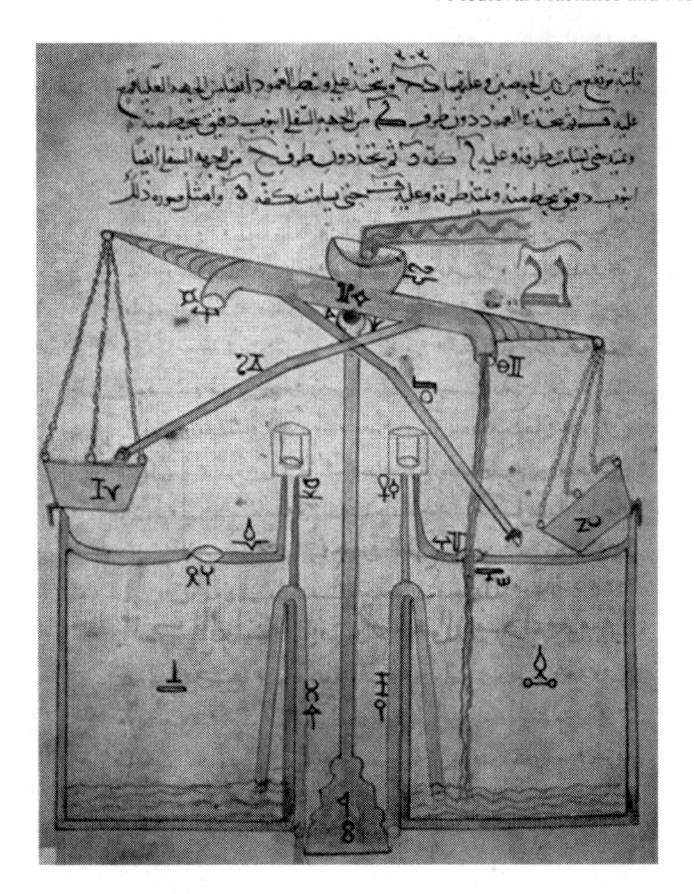

Fig. 4.10 Al-Jazari's fountain [9]

was a thin pipe that emptied on the opposite side into a bucket suspended from the top part. The whole of this machine was pivoted and could tilt from side to side.

In the configuration that is represented in the drawing, the pipe is emptying into the right-hand tank, while water is flowing gently into the left-hand bucket from one of the thin pipes. When the bucket is full, the weight of the water makes the top assembly tilt to the left and it starts to empty into the other tank through different sizes and groups of streams. The cycle is repeated while the water supply is maintained.

On Clocks and Automatons

Around 975, Al Biruni devoted himself to trigonometry, mechanics, and astronomy, and he described the so-called "Moon Box" in his book "An Elementary Treatise on the Art of Astrology".

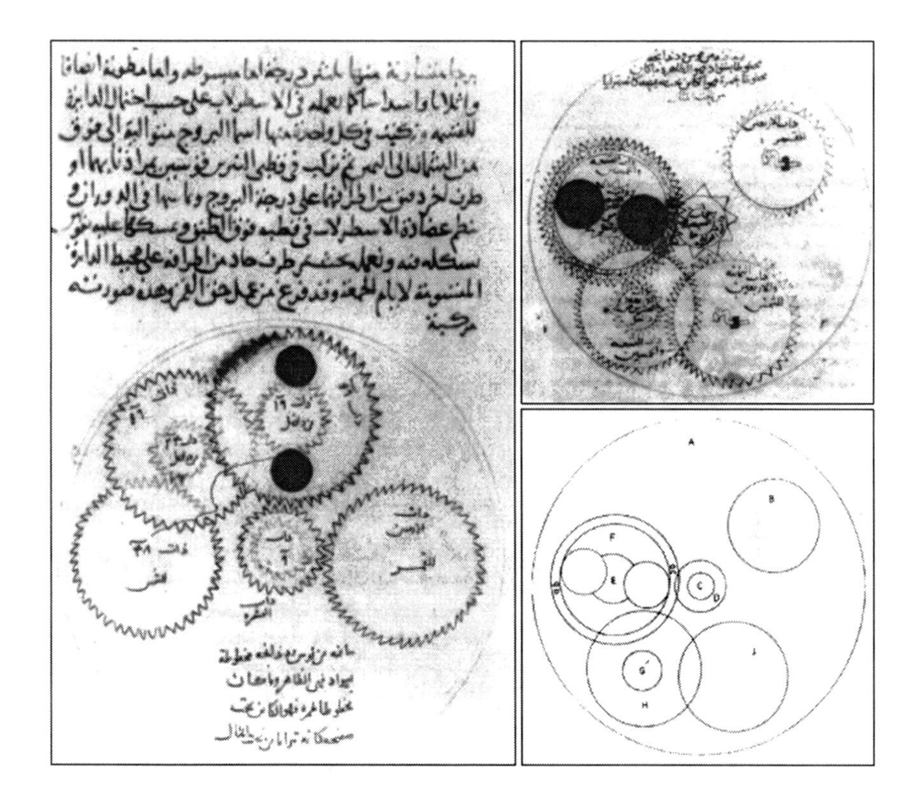

Fig. 4.11 Al-Biruni's Moon Box [4]

The purpose of the box was not only to make an approximate measurement of the phases of the moon but also to measure its position and that of the Sun. It also showed the signs of the zodiac, the days of the week, and the hours. The mechanism was based on eight interlocking gearwheels (Fig. 4.11) each of which had the exact number of teeth for its specific task.

Al Jazari combined clocks and automatons with superb precision. Figure 4.12 shows a candle clock that was designed to measure the passing of 14 equal periods of time. A uniform cross-section candle was used with a wick and a specified amount of wax. A long jacket with a perforated lid was welded to the candle support. A support plate was placed at the lower part of the candle to which a U-shaped channel was welded as divided into 14 compartments where the 14 balls were situated. A weight was positioned to continually push the candle upwards by using a rope-and-pulley system. When the flame was lit, the lid had to be frequently cleaned to ensure the flame remained constant. As the candle was gradually burned, it was slowly forced upwards by the action of the weight and the support plate. At the same time, the pulley system operated an indicator to move. After a time cycle, the first of the balls is loaded into the conduit and reaching the exit, rolls into the bird's head and comes out of its articulated beak to be recovered from a bowl.

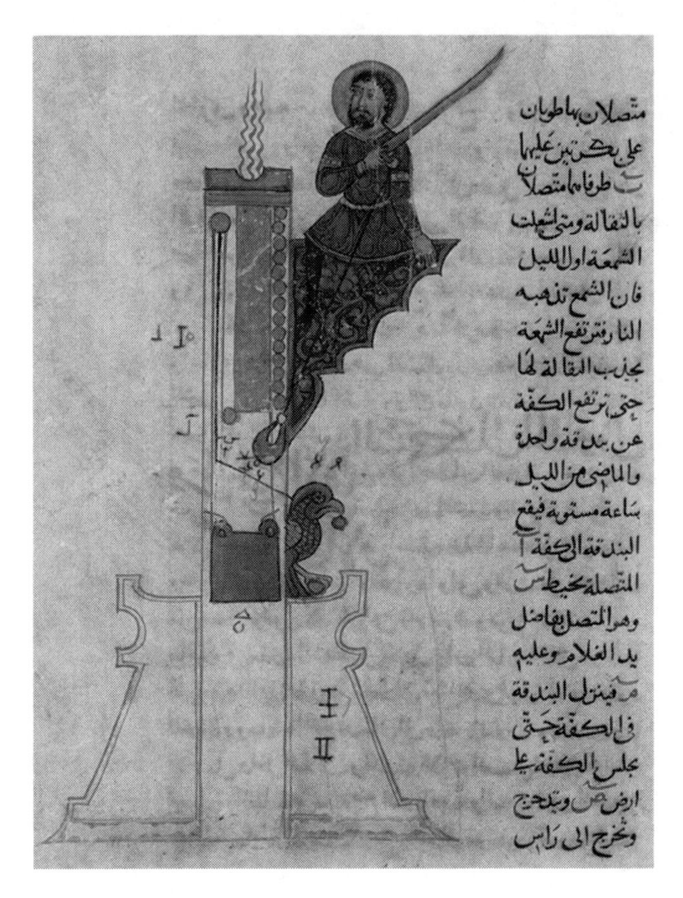

Fig. 4.12 Al-Jazari's candle clock [9]

Figure 4.13 shows a scribe with a quilt in his hand sitting on a bucket-shaped structure. The hour is indicated by means of a horizontal ring set on the structure and divided into 217 parts, which each 15 divisions represent 1 h. The scribe marks the hour with his quilt which is at the beginning of the day in the first division.

For the rotation of the scribe, there is a mechanism inside the tank which can be noted at the side in the diagram. The scribe is connected to a shaft attached to the bottom of the tank and is moved by a pulley that is observable at the foot of the scribe.

When the tank is filled up to the required height, a hole of the right size at the bottom permits sufficient water to drain that, in 1 h, turns the scribe for the 15 divisions. The water flow produces a turning motion through a pulley where a rope is attached to a weight floating on the water. The other end of the rope is connected to a counterweight to retain the weight balance once the rotation is achieved.

When the water has been poured in, the counterweight is at its lowest point but, as the water drains, the lower weight rises and the pulley turns. Thus, the scribe

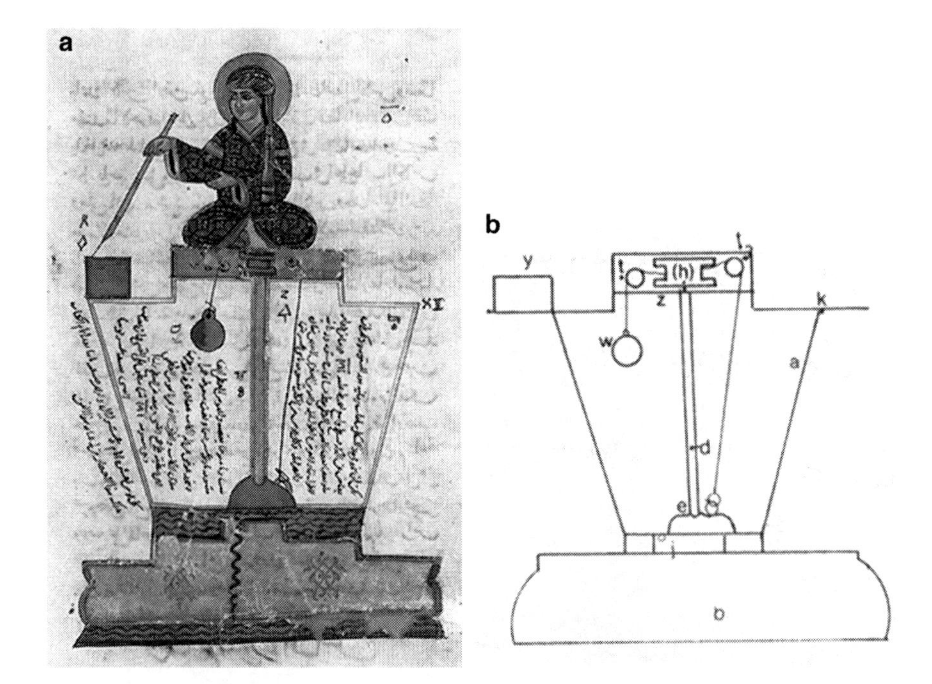

Fig. 4.13 The scribe clock automaton by Al-Jazari: (**a**) the original drawing; (**b**) a reconstruction drawing by Donald Hill [9]

turns with it while the quilt points to the hours. Once the water has completely drained, the tank needs to be filled again.

Of all Al-Jazari's clocks, one of the most famous constructions was the elephant clock that is shown in Fig. 4.14. This was used as an astronomical instrument for the exact measurement of time.

The complexity of this clock design and other similar ones required meticulous assembly. Donald R. Hill showed the mechanisms of some of these clocks in detail (like in Fig. 4.15) when he translated Al-Jazari's book into English in 1979, and made drawings to explain their working.

The bowl (a) floats on the surface of the water in a tank (n), to which it is connected by a joint with several articulations (b) that are indicated on the left. In the upper part of the clock, there is a domed castle that is supported on four columns. Inside the castle there is a ball dispenser that is not shown, from which a conduit leads to a bird head (f). The serpent tail, which in reality is a pulley, is part of a shaft that is installed on bearings. A chain (d) connects the underside of the bowl to a serpent tail, while a cable (h) is connected to the bowl and the ball dispenser by a small piston and a hole (k).

At the beginning of the time period, the empty bowl is on the surface of the water. A calibrated hole regulates the water flow, slowly sinking the bowl until the end of the period when it suddenly submerges. This causes the cable (h) to operate the ball

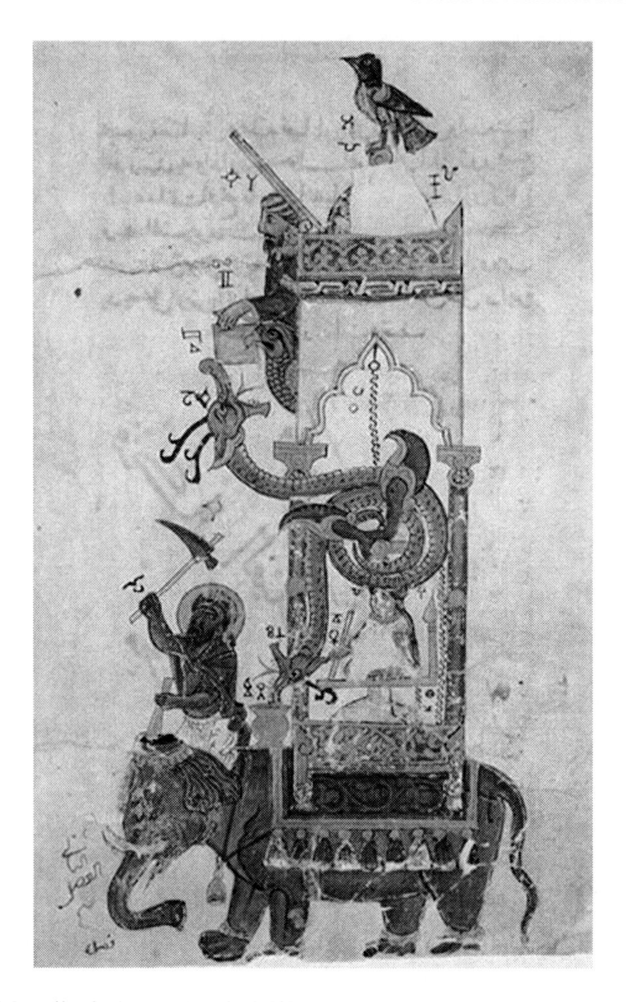

Fig. 4.14 Al-Jazari's elephant water clock [9]

dispenser and a ball falls from the bird beak into the serpent's mouth. The serpent's head drops and the chain (d) pulls on the bowl, which empties its contents since it is articulated at (b).

The ball drops from the serpent's mouth and strikes a small bell. When the whole movement is finished, the serpent's head returns to its initial position. The empty bowl is again horizontally floating on the surface of the water and the cycle starts again. The clock continues to work while there are balls in the dispenser.

The water flow regulator can be considered to be one of Al-Jazari's great contributions as it consisted in a perfectly calibrated hole through which the bowl gradually submerged to produce the exact flow velocity for different variations in water velocity. It was this immersion that marked the time of the hours, and this means

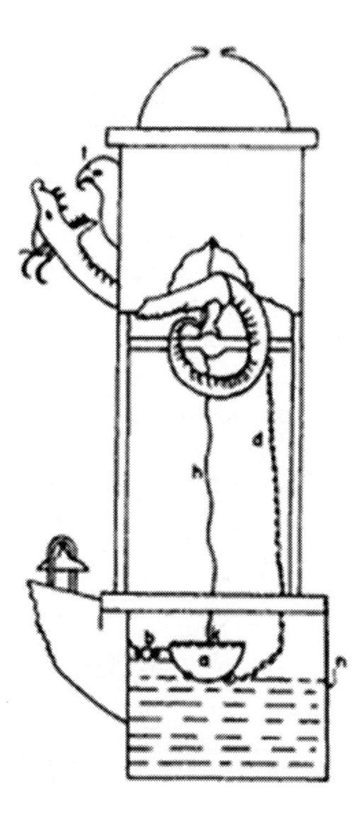

Fig. 4.15 Donald R. Hill's drawing for explaining a Al-Jazari water clock [9]

that Al-Jazari must have performed several experiments and trials before coming up with the exact size of hole to obtain a perfect hour counter.

Al-Jazari used the force of gravity as an engine to make the bowl sink and also for the dropping movement of the serpent's head when it had a ball in it.

Besides these mechanisms, he used two others, namely a return mechanism and a control mechanism. As already noted, the return mechanism is activated when the ball has dropped from the serpent's head, while a pulley makes the head return to its original position, and the submergible bowl rises to the surface, losing the water that it has picked up.

The control mechanism is located in the bowl and the control law is marked by its fall-and-rise cycle, which is maintained while there are metal balls in the dispenser (as a closed loop cycle). It can be noted that a cable and chain are attached to the bowl; the cable goes from the ball to the bowl, and it is this that releases the mechanism inside the castle and activates it when the bowl has sunk. However, the chain goes from the underside of the bowl to the serpent's tail and its task is to tilt the bowl to empty out the water.

One example of the design mechanical complexity and precision was a clock that gave information on the phases of the Moon and the position of the Sun in the signs of the zodiac. This was a measurement of time not only on an hourly and daily bases, but also on a monthly and yearly basis.

The system in Fig. 4.16 shows this complex mechanism, which has a wheel at the top that is illustrated with the 12 signs of the zodiac. Below there are two half-circumferences; the upper one marks the state of the Sun by means of a golden sphere and the bottom one marks the state of the Moon by means of a glass sphere.

The appearance of a figure at each of the 12 windows on the top row marked the passing of the hours. The row of doors beneath changed colour depending on whether the indicated hour was for day or night. At the same time, the two birds at the sides tilted forward by pushing a sphere to fall from their beaks into the goblets and by activating some cymbal sounds. In addition, at the sixth, ninth and twelfth hours, a device was pressed that activates a music band.

The clock ran on a complex system that is based on a water-flow regulator to measure the times, as shown in Fig. 4.17.

The movement was carried out by the lower guide along which a carriage moved horizontally. Consequently, this carriage moved the figures appearing in the windows of the upper frieze.

Fig. 4.16 Al-Jazari's clock [9]

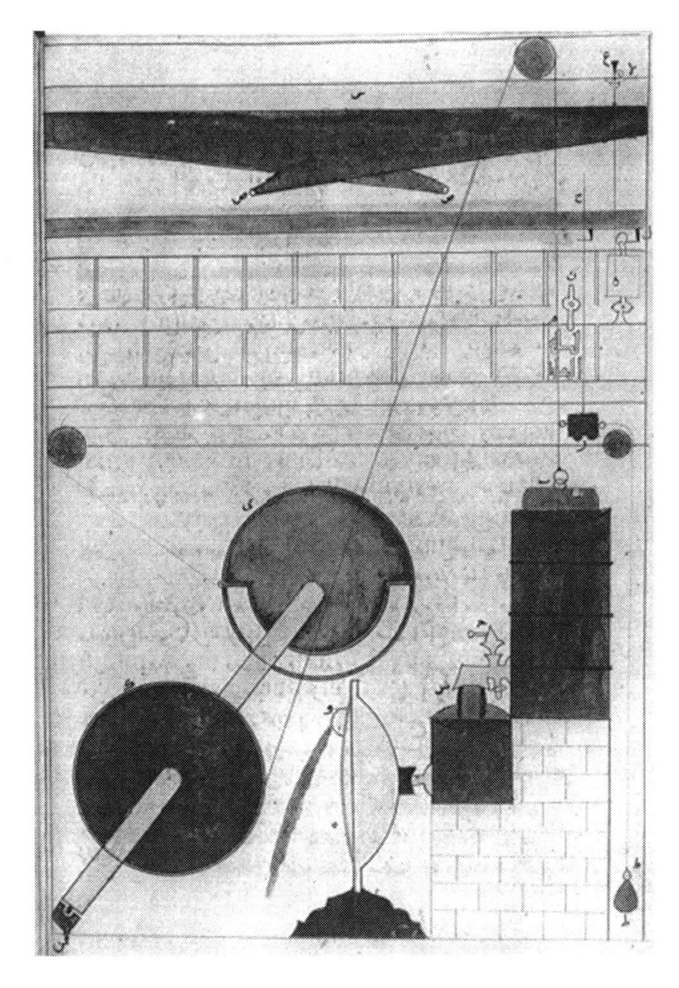

Fig. 4.17 Time marking mechanism [9]

The system was based on a water tank whose water flow was regulated by a tap and a series of pulleys and pistons which transmitted also the other movements. The water coming from the tap moves continuously down the piston, which pulled on the upper pulley and the large lower wheel. As this wheel is rotated, the shaft on which it is installed, is rotated and, consequently, the upper wheel is rotated too. The thread connecting this wheel and the carriage guide completed the movement.

The water regulating mechanism had to be perfectly calculated in order to stop the water falling after 12 h, and, consequently, the top of the shaft with the carriage that is located at the last window corresponding to the twelfth hour. When the tank was again filled, the shaft rotated in the opposite direction and the carriage returned to the first hour in order to be ready to begin the cycle again.

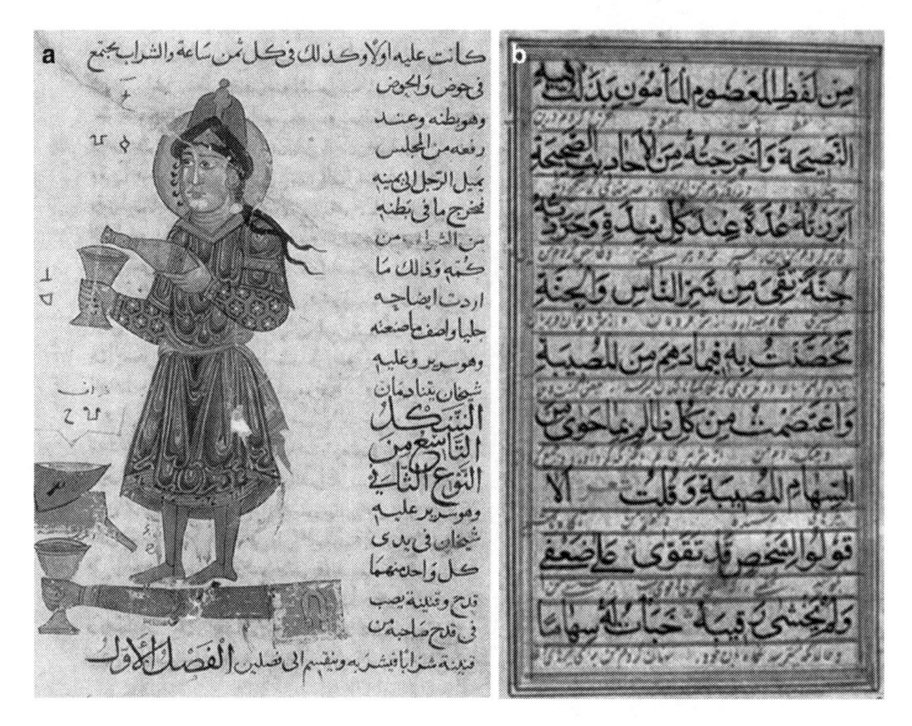

Fig. 4.18 Pages from Al-Jazari's book. (**a**) Drawing and explanation of an automaton. (**b**) Page of explanations [9]

Figure 4.18 shows two pages from Al-Jazari's book illustrating details that he wished to present his work as fully descriptive and detailed drawings and precise explanations.

Al-Jazari presents many more automaton mechanisms for several purposes, such as to serve wine, to dispense fruit, for washing hands or making music. These automata were used as toys or for entertaining guests.

The illustrations in Figs. 4.19 and 4.20 show some of these automata.

All these are prime examples of a refined culture that devoted a large part of its mechanical brilliance and inventiveness to producing luxury objects endowed with movement. The overwhelming superiority of Muslim trade over many centuries should not be forgotten. Their trade relations were extended to the whole of the known world, and particularly products that were directed to the upper classes of the cultures they had relationships with. Automatons, like those described above, would have been among those products.

Prior to the authors previously mentioned, in the ninth century the Banu Musa brothers wrote their "Treatise on Ingenious Devices" which had a great influence on subsequent machine design in the Muslim world. Although it can be considered a key book for the history of machines, the explanations given in it and the drawings themselves are not fully clear. This sometimes makes it difficult to interpret the illustrated machines, which have been clarified by the comments of later Arab writers.

Fig. 4.19 Al-Jazari's wine and water serving devices [9]

Fig. 4.20 Al-Jazari's automatons [9]

Nevertheless, in order to complete the contribution of Muslim culture, this history cannot miss to mention the Banu Musa machines. Figure 4.21a shows a lamp installed inside a hemisphere that is capable of self-adjustment by means of a rack and pinion. Figure 4.21b shows two of the Banu Musa brothers' even more advanced devices. As the devices evolved water energy was used as the driving force.

They are at least examples of how mechanical automatons were developed, from the beginning of the spread of Islam.

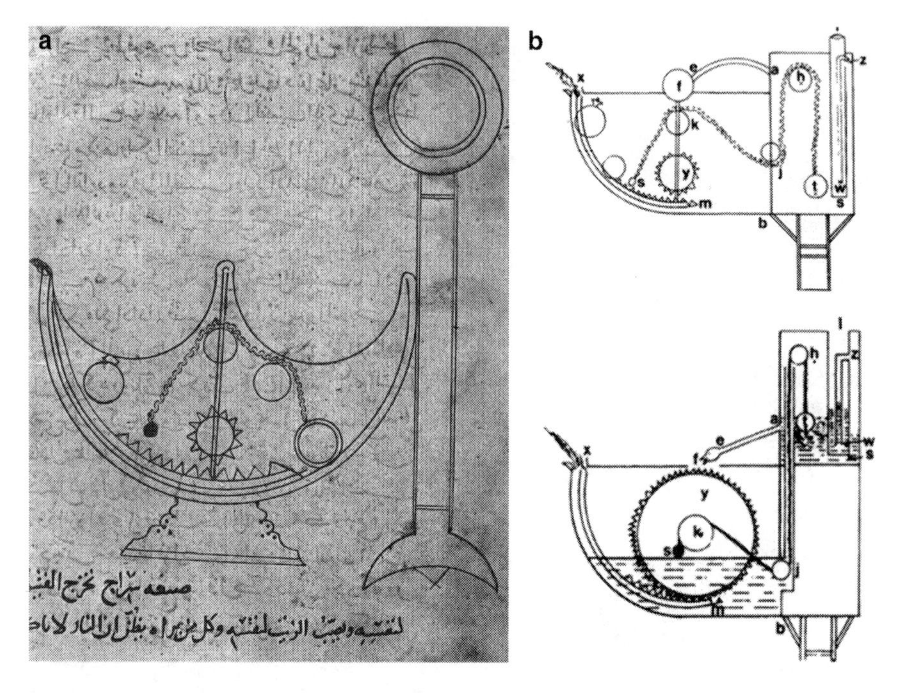

Fig. 4.21 One of the Banu Musas' lamps: (**a**) the original drawing; (**b**) Lamp diagrams for interpretation [63]

On the Transition in Europe

Progress in the Arab and Asian worlds did not appear to have reached Western Europe until the thirteenth century. However, some European inventions did exist, like the arrow launcher shown in Fig. 4.22, that is taken from a drawing by Villard de Honnecourt with a similar structure to some of the Chinese catapults in Chapter 2.

With the help of Fig. 4.23, the way the machine works can be clearly explained thanks to the interpretation through drawings by the French National Library (BNF). The first thing to surprise us is the sheer size of the device if we consider that the figures are drawn in scaled size, since Villard de Honnecourt's machine was apparently of 18 m high.

The mechanism consisted in driving the arrows forward by using the thrust of a pivoted plank. This plank was made to fall with the help of two pulley systems that were on each side on the ground and were attached to ropes also connected to the plank. The force required by the operators was not only to drop the board but also to raise the attached counterweight. Villard defines the counterweight as "an enormous basket full of two large "toesas" of earth (French unit equal to 1,949 m long, 9 ft wide and 12 ft deep, once the counterweight had been raised, the rope was cut (Fig. 4.23b) transmitting the movement of the counterweight to the arrows when the

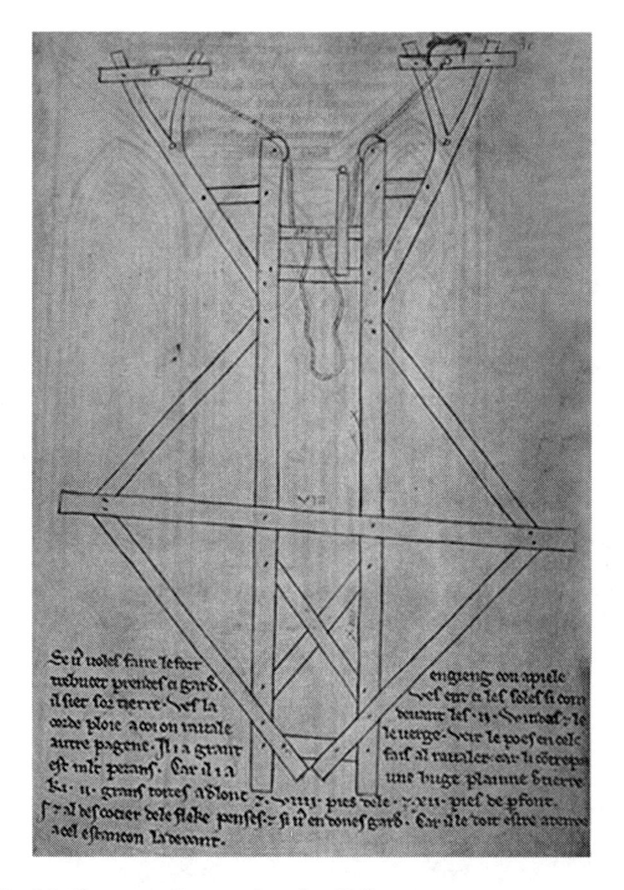

Fig. 4.22 Villard de Honnecourt's arrow launcher [64]

edge of the plank hit them. Obviously, a machine of such a size provided a large force and obtained very long shots that could have been useful for besieging cities or fortresses. The studies made by the French National Library refer to 100 kg projectiles whose energy necessary was able to destroy bridges or smash through defence walls.

The book begins with the words "Villard de Honnecourt greets you and asks that all who use the devices in this book pray for his soul and remember him. For this book shall be of great assistance in building work and in joinery machines ..." One of those joinery machines is a saw for cutting piles in water, as illustrated in Fig. 4.24. Both the original figure and the one by the French National Library show the peculiarities of the mechanism.

The saw in the drawing is fixed horizontally to a frame situated above the water and it is supported on a platform, where two workers on either side of the saw, as shown in the figure to the right, move the platform by pushing it backwards and forwards.

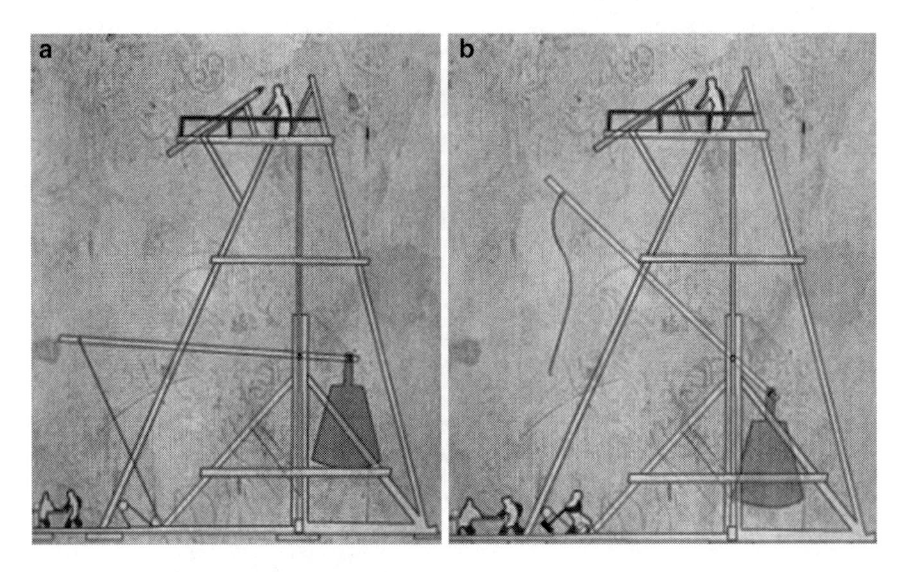

Fig. 4.23 Positions of Villard de Honnecourt's arrow launcher. (**a**) Falling. (**b**) Rising. Reconstructions by the BNF

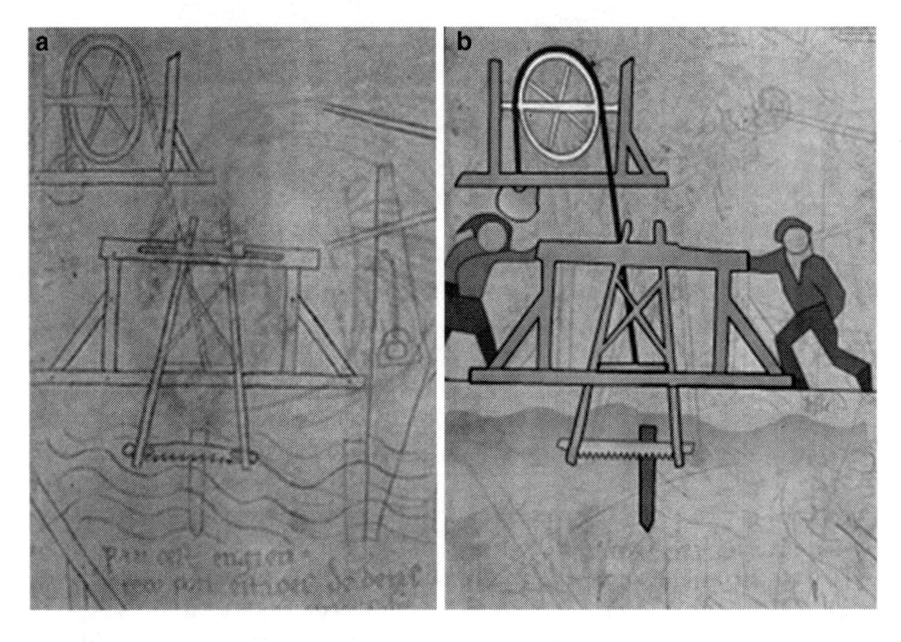

Fig. 4.24 Villard de Honnecourt's saw for cutting wood under water. (**a**) Original drawing [64]. (**b**) Reconstruction by the BNF

Villard de Honnecourt placed a wheel with a counterweight attached to the saw with a rope so that it would exert pressure on the pile to be cut and assist the movement. He also drew a plumb-line to the right.

The book also contains a water-driven saw. Honnecourt writes of this machine: "A sword is thus made that saws all by itself". Although there is a lack of illustrated documents, everything would seem to point to the existence of hydraulic power in the West, probably for fulling mills and other uses.

Figure 4.25a is the first illustration of a hydraulic saw. It had previously been described and used but never drawn until Villard produced his notebook. If the crudeness of the actual drawing reflects the actual construction, the difference from contemporary designs from the Muslim world is highly remarkable.

In this saw, the circular motion of the waterwheel creates an alternate rising and falling movement that is capable of sawing wood, to which is added a wheel's automatic forward movement towards the saw. The water turns the wheel by means of the schematically represented paddles and, consequently, its shaft rotates the wheel with four cams. A drag wheel is used to advance the piece of wood that is held among four supports to stop it moving from the horizontal position. The work of the cams is to drive the articulated arms at the foot of the saw.

This second movement is based on the attachment of the saw at the top to a flexible pole. When the articulated arm is leaned on, the cam forces the saw down, which bends the flexible pole and then it makes it rise again to its original position. This is an impulsive movement but it is effective, since the lower articulation was designed so that the movement does not lose its verticality.

The saw guide mechanism may be interpreted as quadrilateral where a coupler is used to guide the saw in its alternate movement. This mechanism was not to be used again until James Watt's steam engine appeared in 1775. By observing the machine in Fig. 4.25, we can conclude that technical development took place during the Middle Ages in Europe. However, there was a limited awareness and spread of this technical culture which only reached maturity during the Renaissance.

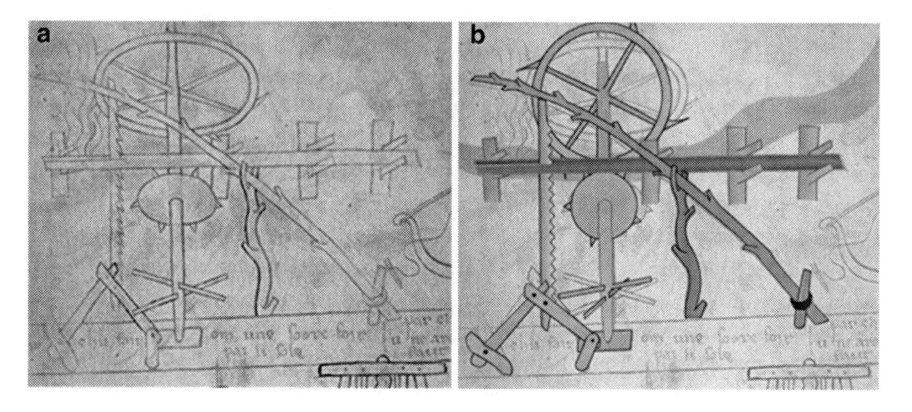

Fig. 4.25 Villard de Honnecourt's hydraulic saw. (**a**) Original drawing [64]; (**b**) a reconstruction by the BNF

Fig. 4.26 Villard de Honnecourt's elevator [64]

Thanks to Villard de Honnecourt, we also have an illustration of an elevator as shown in Fig. 4.26.

The reconstruction in Fig. 4.27 depicts a wooden shaft where the top two-thirds have been turned, while a horizontal handle has been added to the bottom third to produce the rotary movement of the shaft. The ends of the screw are fixed, while a nut and some crossbars stop it turning. A load-bearing rope is attached to the nut. In the BNF's reconstruction, a set of pulleys has been included to avoid contact between the raised component and the upright shaft. The upright shaft is turned by manpower by moving the nut to raise or lower the load.

The optimism of the time led some writers to design impossible machines that had perpetual motion and could keep moving permanently without any external energy input. Villard de Honnecourt was not indifferent to this tendency and drew one of the first designs of this type of machine, as shown in Fig. 4.28.

The device in Fig. 4.28 used hammers that, once in motion, would receive sufficient impulse from gravity to keep the wheel in perpetual motion. Underneath the drawing, Villard de Honnecourt wrote: "For some time experts have been discussing how to make the wheel turn by itself. This may be achieved by an odd number of small hammers and mercury in the following way".

After these examples, it is evident that the name "Renaissance forerunner" is more than deserved by this architect-engineer.

Fig. 4.27 Illustration of Villard de Honnecourt's elevator. Reconstruction by the BNF

Fig. 4.28 Villard de Honnecourt's perpetual motion device [64]

Chapter 5
The Machine Renaissance

The Renaissance in Western Europe in the fourteenth to sixteenth centuries marked strong activity of recovery and revitalisation in artistic, scientific, and literature frames that overcame the stagnation of the Middle Ages. Unlike during medieval times, the opening of society during the Renaissance promoted the spread of machines in many environments. Little by little, many existing machines were no longer considered simply as a means of carrying out civil or military engineering works.

The first modern approaches to machines in the fourteenth century opened the way to the activities of the fifteenth century, which endowed machine technology and raised the engineering profession to a status with some dignity. Thus, since the sixteenth century recognition was given to machine experts and then, since the beginning of the seventeenth century, machine knowledge was treated as an academic subject.

During the Machine Renaissance (which is the Renaissance period in which there was strong reconsideration and evaluation of machines), two separate but related activities may be recognised; namely they are theoretical study with a scientific approach and professional practice of an experimental nature. Both lines gradually converged until they came together in the seventeenth century.

The fifteenth century might be considered the high point in machine development, with outstanding personalities such as Francesco di Giorgio (1439–1501) and Leonardo da Vinci (1452–1519). At the same time, an interest in the theoretical aspects of machines led to the knowledge from Antiquity being recovered. Greek mechanical works were again examined and the machines of Roman engineers were studied and copied. These works were translated, interpreted, and illustrated, since the original illustrations had never existed or had not been preserved.

Leonardo da Vinci, of widespread fame and a brilliant mind, was able to develop machines with a high level of genius because he lived in an environment that accepted his ideas. He also faced a lot of competition from other designers and builders of similar technical ability. Indeed, Francesco di Giorgio may be recognised as a greater engineering genius with tremendous innovative skills and a wide-ranging production of machines.

The publication of machine knowledge in the form of treatises began at the end of the fifteenth century. A first line of activity was the study of machine mechanics

E.B. Paz et al., *A Brief Illustrated History of Machines and Mechanisms*,
History of Mechanism and Machine Science 10, DOI 10.1007/978-90-481-2512-8_5,

as an application of physics that was undertaken by outstanding personalities like Guidobaldo del Monte and Galileo Galilei (see Chapter 8). The second line consisted of a development towards a discipline in the form of the rational collection of machines. Outstanding were the machine collections of the afore-mentioned Francesco di Giorgio and Agostino Ramelli (1588).

The Machine Renaissance spread from Italy to all of Western Europe from the second half of the fifteenth century, with milestone works, examples of which are "De Re Metallica" by Georgius Agricola (1494–1555) and "Kunstliche Abris allerhand Wasser" by Jacobus Strada (1515–1588).

New solutions were also sought for new applications in order to increase machine capacity and precision and to adapt them to specific circumstances. Filippo Brunelleschi (1377–1446), an architect and draughtsman, is worthy of mention in this respect, since he was a ingenious machine designer for his architectural goals. In general, there was a wide variety of machines for very different applications and with ever-increasing complexity. Work was mainly undertaken in three areas: hydraulics, lifting devices, and war machines, while other devices such as clocks and automatons were left in the background.

The illustrations in this chapter are a limited selection of figures that can be found in machine works of the period. Figure 5.1 shows the title pages of examples of these "Machine Treatises", namely "Le Machine" by Giovanni Branca from 1629, "Theatrum machinarum novum" by Georg Böckler from 1661, and the "Novo teatro di machine et edificii" by Vittorio Zonca from 1607.

As regarding war machines, it was Leonardo Da Vinci (Atlantic Codex, 1500) who produced the most amazing sketches. Among the many later authors, we should not forget Heinrich Zeising ("Theatri machinarum erster") and George Andreas Böckler ("Theatrum Machinarum Novum").

Some of the authors of the period were so brilliant that they eclipsed other great engineers who built both new and old machines based on earlier designs. There are also great works by anonymous authors who have failed to reach recognition, almost certainly because their notes and writings were not preserved across the centuries.

The printing press was an important factor in the dissemination of these treatises. Not only the text but also the accompanying engraved printed figures reached a quality that was previously unknown in books on machines that were patiently copied by scribes, who frequently had no mechanical training and made reproductions that were erroneous, if not confusing. Although some significant treatises have reached present times in the form of manuscripts, most authors published printed books whose readers no longer needed to belong to the privileged classes of high culture. Since then, professionals had access to mechanical engineering knowledge without depending on the libraries of the powerful mecenates. Machine knowledge became popular and spread on a scale that was different in quality from all that had occurred before.

The intellectual curiosity that was developed form the Renaissance spirit, and then the printing press turned the history of machines into something fundamentally European.

Fig. 5.1 Title pages of some "Machine Treatises" [24–26]

On War Machines

The notebooks of Leonardo Da Vinci (1452–1519) in the "Atlantic Codex" and other manuscripts show several war machines that were designed around 1490. A particularly outstanding model is reported in Fig. 5.2 showing a scythe chariot that was designed to prevent the enemy getting near it.

It is interesting to see how simplicity of design can create a powerful weapon driven by two horses. The chariot wheels have radial hooks to grip the ground and, simultaneously, they start a series of gears and transmission shafts until the movement reaches a rear blade and a front rotary cross-blade. Thus, a defence system is formed on all four sides of the machine.

The catapult is a war machine that had been well known since Antiquity, as we have seen in the chapter devoted to China (Chapter 2), but it was also studied in-depth by Leonardo. Its new design was not based so much on changing the aesthetics of the device, but on changing its mechanics. His study on springs led him to design a new catapult with a better performance than those built up to that time.

According to Fig. 5.3, when the spring is loaded, the two wooden arms bend towards each other, generating tension and operating a crank that takes up the rope. When the device is ready, the handle is released to launch the projectile on the spring. What made the catapult more effective and suited to long-range launches was the stiffness of the springs. Moreover, Leonardo designed a system that made it easy to load the projectile without excessive force by designing a lever operating the spring movement without the side rods and spokes impairing that movement.

Fig. 5.2 Scythe chariot from the "Atlantic Codex" [132]

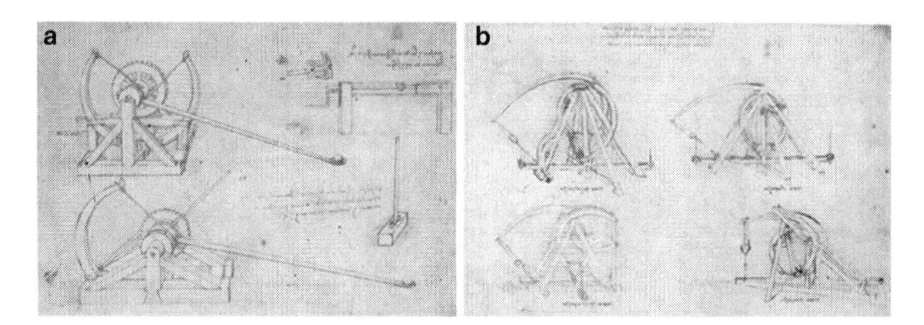

Fig. 5.3 Catapults, from the "Atlantic Codex" [132]

On Lifting Machines

Filippo Brunelleschi elaborated relevant developments in lifting/crane machines for architectural work, such as for building the dome of Florence Cathedral. All his inventions reached us through drawings by other designers, who referred to his sketches and designs. These descriptions were hand notes for personal use and for building those systems through their own pupils/co-workers. With this character they circulated confidentially among professionals for machine design of the time.

Figure 5.4 shows the core mechanical transmission for a light crane machine by Brunelleschi from a drawing by Bonaccorso Ghiberti (1451–1516). The movement is achieved through two crossbars that are located at the top of the drawing and joined to a lantern that needs four men to operate them. The horizontal lantern engages a vertical geared wheel that drives a transmission axle on which a pulley is also installed.

Figure 5.5a is also based on Brunelleschi's design and was drawn by Mariano di Jacopo, who was better known as "Il Taccola" (1382–1458), and was a machine designer highly appreciated by his successors. In his practical machine studies he conceived outstanding solutions for several applications.

Mariano di Jacopo may be considered as one of the first professionals whose main activity was focused on machine development.

Figure 5.5a shows an animal-powered lifting system. The size of the machine can be seen by comparison with the dimension of the horse. The horse turns the upright shaft to which a screw is attached and two other horizontal gearwheels move accordingly. As both wheels rotate, they transmit the rotary motion to a vertical gearwheel that operates a rope-and-pulley device. Since the screw nut is fixed to the base, the horse's movement also causes the upright shaft to rise. Brunelleschi used this system to improve lifting tasks.

Other crane machines were presented by Giuliano da Sangallo (1443–1516), together with details of mechanism designs. Figure 5.5b shows a machine design, which is also based on the use of one for lifting a load and another screw for radial movement. The mechanism design allows a complete rotation of the crane structure too.

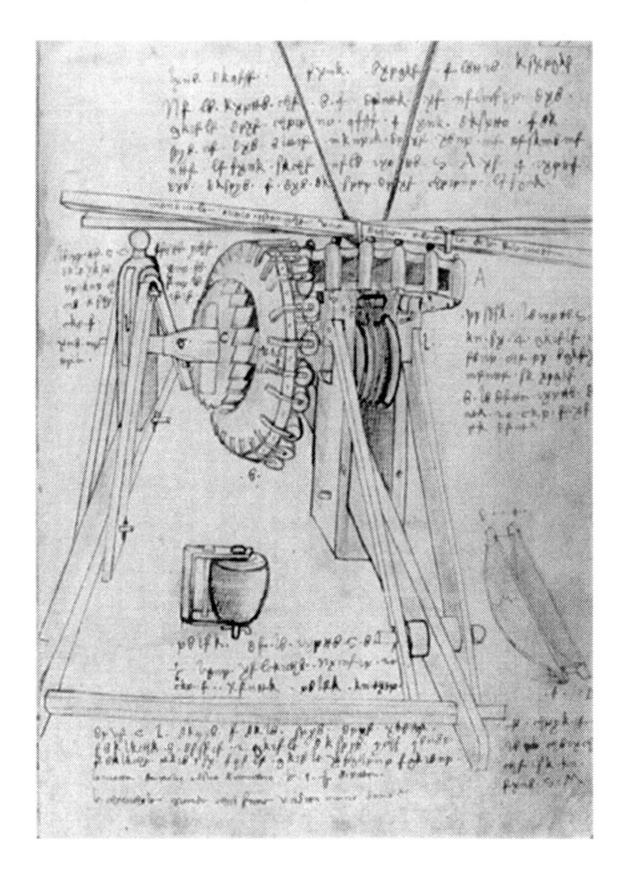

Fig. 5.4 Mechanical transmission in a light crane machine by Brunelleschi from a drawing by Bonaccorso Ghiberti

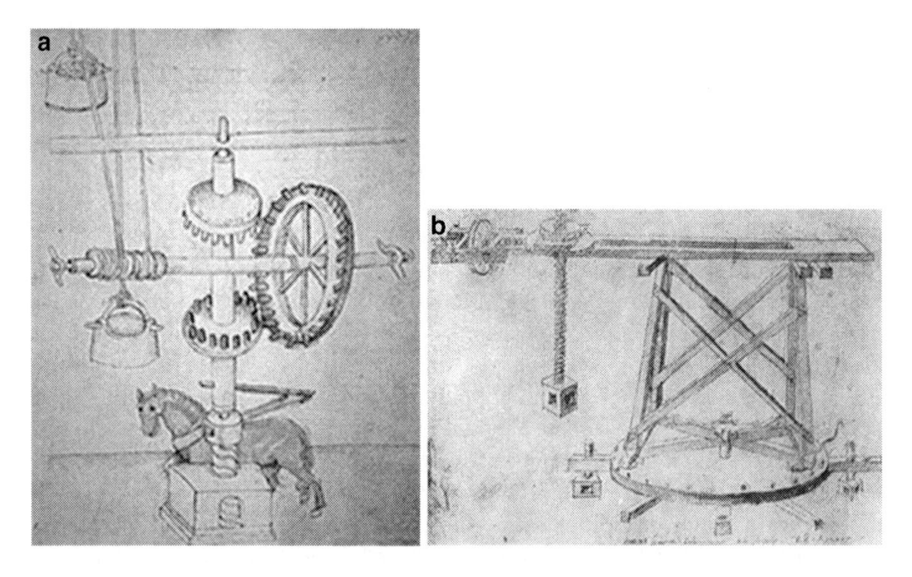

Fig. 5.5 (**a**) Horse-powered crane by Mariano "Il Taccola". (**b**) Crane by Giuliano da Sangallo

Load lifting, radial movement, and load rotation movements could be carried out independently.

Figure 5.6 shows two drawings of the same machine. Figure 5.6a is a drawing by Bonaccorso Ghiberti, and Fig. 5.6b was made by Leonardo da Vinci. Both drawings describe a rotary crane that was designed by Brunelleschi. When considered together, they clearly show how the machine worked. The upright shaft in this drawing is made to rotate by a lateral guide to place the load as it rises. Both the load and counterweight can be moved radially to keep the machine's equilibrium. The load moves vertically through the action of a worm screw. Four men were needed to operate the crane: two men to turn the structure, two others to turn the screws for the radial movement of the load and the counterweight and a fifth man is needed to operate the vertical screw.

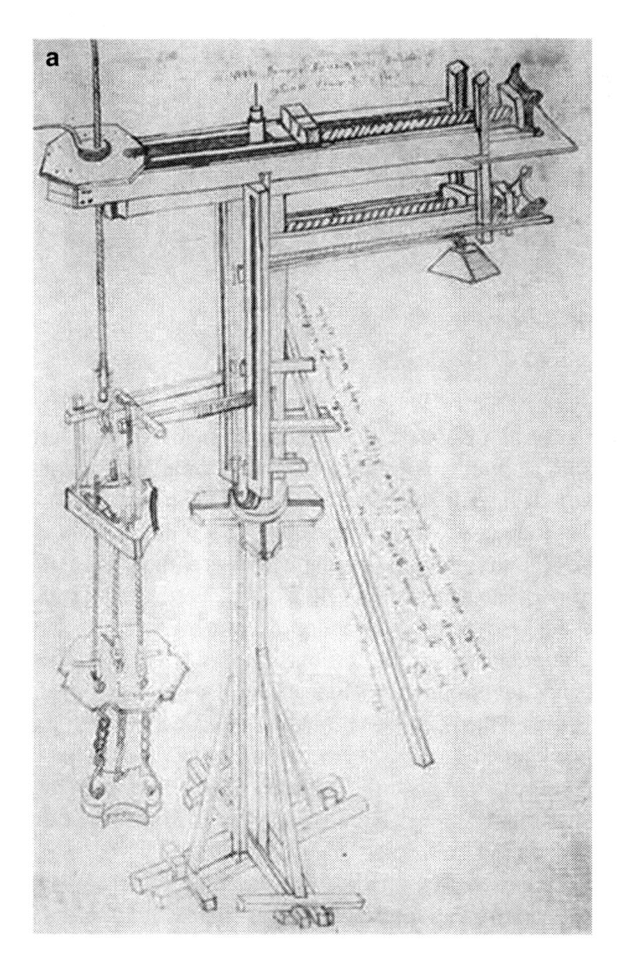

Fig. 5.6 Drawings for a rotary crane of Brunelleschi, by: (**a**) Bonaccorso Ghiberti,

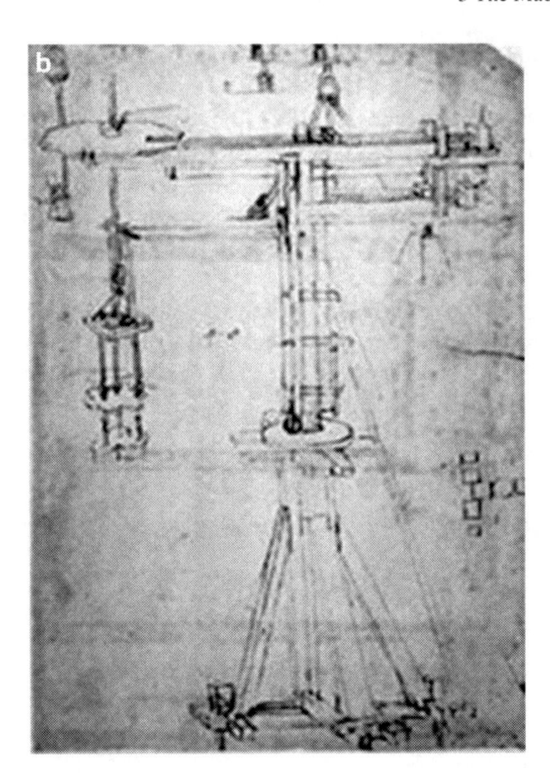

Fig. 5.6 (continued) (**b**) Leonardo da Vinci

Among the several machines that were designed by Francesco Di Giorgio (1439–1501), with a simultaneous combination of simplicity and genius, there are sketches of lifting devices like the one in Fig. 5.7. The mechanism is divided into two parts. At the bottom, a winch is operated by a train of gears and pulleys that are located at the bottom right, terminating with two pulleys, each of which controls a hook. The top part has two screws which allow vertical and radial movements, respectively, of the load that is suspended from a hook. It can be noted that the upper part of the crane can rotate around its upright shaft. This machine can be understood as a development of Brunelleschi's previous machine (Fig. 5.6). Di Giorgio incorporated the principles behind its operation, but with greater simplicity, by adding an independent dual hook mechanism for additional load lifting.

In the later designs by Agostino Ramelli (1588), each design is described in more detail by describing how it is built and how it works. An example of this is the lifting machine in Fig. 5.8.

Ramelli added the capacity to manoeuvre the lifted object by rotating the whole crane frame by operating a wheel at its base. Thick wooden rods were used to strengthen the central structure and an inclined beam with metal supports were installed to resist the stress that is produced when lifting a load.

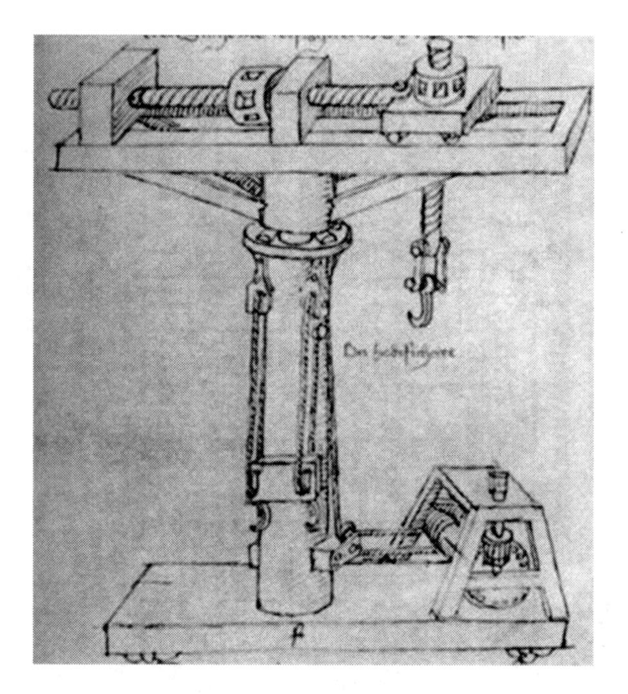

Fig. 5.7 Bridge-crane by Di Giorgio, shown in the "Trattato di architectura e machine" [52]

Fig. 5.8 Lifting mechanism, from Ramelli's "Le Diverse et Artificiose Machine" [93]

Fig. 5.9 Zeising's lifting mechanism from "Theatri machinarum erster"

Figure 5.9 shows one of the machines of Heinrich Zeising (1611), which is very similar to the previous one. This is an example of diffusion in European frames thanks to the exchange of ideas and knowledge. This influence became more clear after 1580 when European books included illustrations of very similar machines that occasionally showed some innovation.

Around 1400, the works of Mariano di Jacopo, "Il Taccola", were published in his book entitled "De Ingenis". He not only studied the designs of his predecessors, as we have seen with Brunelleschi, but he also conceived new ideas by studying the mechanisms for deployable stairs.

The mechanism in Fig. 5.10a consists of three connected stairs, where the stair height is increased by a movement that operates the relative stairs motion by a screw at the bottom. This is an illustration of the use of a screw transmission with a fixed nut and a lever-operated screw.

These mechanisms may be compared to those of Roberto Valturio in his "Re militari" almost 150 years later (1535), as in Fig. 5.10b. They are very good examples of the backwardness in the rest of Europe as compared to Italian mechanical know-how.

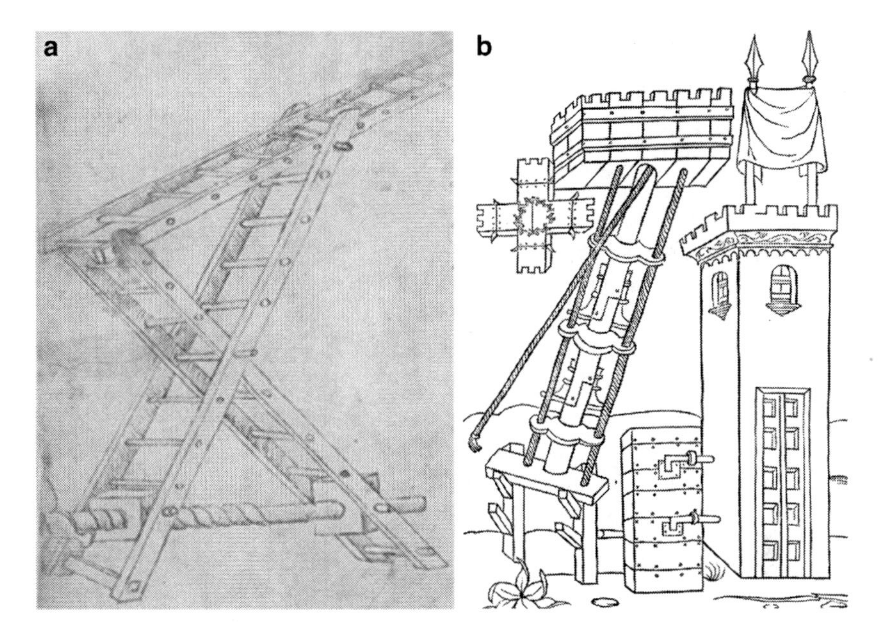

Fig. 5.10 (**a**) Extendible ladder from "De Ingensis" [65]. (**b**) Extendible platform from "Re militari" [123]

On Hydraulic Machines

In his "Trattato di architectura e machine", Francesco Di Giorgio explains the theoretical principles behind hydraulic machines. His designs are mainly based on water-suction machines that direct the water towards channels or buckets.

Figure 5.11 shows part of a page from Di Giorgio's book (see Chapter 8), in which he approaches the problem with a theoretical introduction and then illustrates his design expertise with several machine designs. It should be pointed out that the illustration shows an articulated mechanism of great structural complexity, with modern conceptual design.

Figure 5.12 shows a hand-operated dual-piston pump. A crank moves a pinion that simultaneously engages two vertical racks, making them to move in opposite directions. The vertical movement of the racks regulates the movement of two arms that are joined to the pump pistons. In this way, a more uniform flow is obtained, since the two pumps operate alternatively to raise the water, which flows out through side holes in the pistons.

The mechanics of converting circular motion to vertical rectilinear motion was achieved by using a rack-and-pinion transmission. The handle that operated the pinion had to reverse its direction after each cycle. For obvious mechanical reasons, the crossbars could not be connected to the upright ones. The sliding joint was required between them and it was obtained by using running systems where rollers were placed to facilitate movement between them.

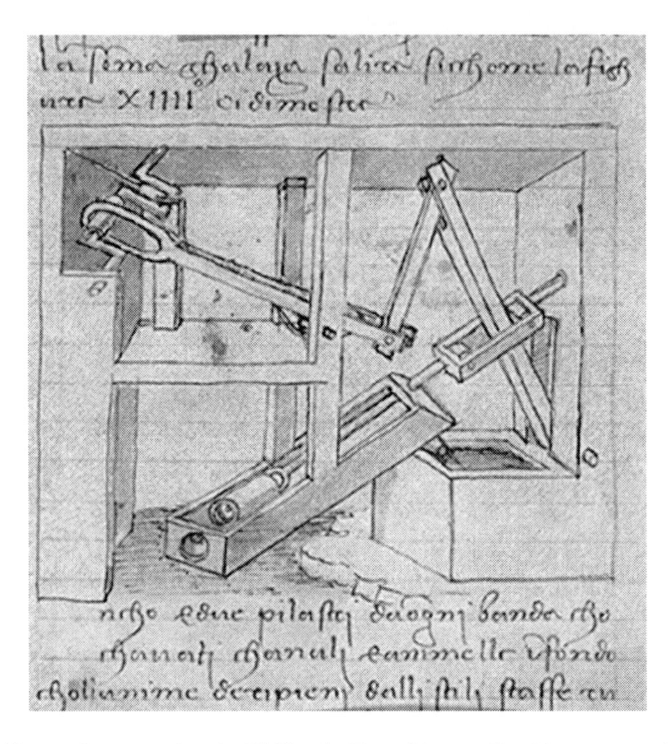

Fig. 5.11 Theory of pumps written by Di Giorgio (From "Trattato di architectura e machine" [52])

The design as a whole has a rather primitive appearance and it was probably not very efficient. However, this demonstrates a broad knowledge of mechanical sources that could be used for making machines.

It is pointed out that the text and illustrations are of an age when manuscripts were still used as a means of dissemination.

A century later, the book"De Re Metallica" (1556) by Georgius Agrícola (1494–1555) was published in which the author reported all known methods for mine work in 12 short chapters that are full of drawings and text. This shows an example of how the rest of Europe echoed the very rapid evolution of the Italian masters.

Agrícola used the printing press for both texts and engravings, which resulted in clear, detailed designs, although the machines have complex mechanism structures. Figure 5.13 is an explanatory drawing combining the perfection of clear visual layout with the utility of a practical machine. The drawing shows a mechanism for sucking water from a well or a mine gallery. It consists of two stages for raising water with a tank at a halfway level.

A river current is used to move a waterwheel that pushes the whole mechanical design into motion. Then, a single gear assembly distributes the movement to two cranks that operate in counter-phase to alternately move the two pistons and to raise the water. Worthy of mention is the articulated link mechanism that is used to move

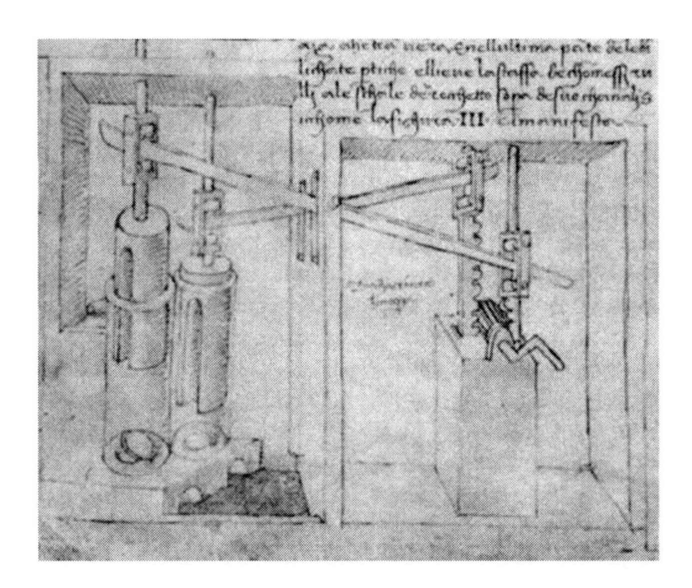

Fig. 5.12 Di Giorgio's water suction machine from "Trattato di architectura e machine" [52]

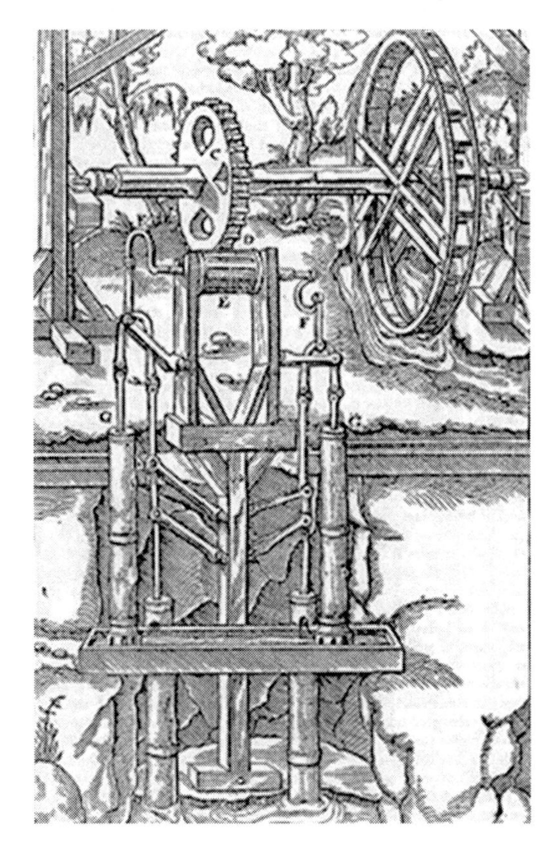

Fig. 5.13 Agricola's water suction mechanism (From "De Re Metallica" [1])

the upper and lower pistons simultaneously, by maintaining the pistons in a vertical position to prevent their blocking inside the cylinders.

Agricola also combined hydraulic force and lifting mechanisms in some of his sketches, as shown in Fig. 5.14. The huge crane is an example of the force that was needed to raise earth from the depths of a mine. In the drawing, we can note that the pulley that is marked with the letter K is moved by the enormous wheel to its right which, in turn, is operated by the hydraulic force of the river. We can also note that the man in the cab (as marked with the letter N) moves two rods that are attached to two brakes that slow down the wheel when the pulley has raised the load and the bucket needs to be emptied. The man and the ladder on the left might simply be part of another even larger mechanism that would also be used for lifting.

Agostino Ramelli, in his "Le Diverse et Artificiose Machine" (1588), designed a large number of suction assemblies like the one in Fig. 5.15. It is a development of previous designs incorporating worm gears and rack-pinion transmissions, by using parallel suction pumps.

Fig. 5.14 Agricola's lifting mechanism (From "De Re Metallica" [1])

In the case of Fig. 5.15, the machine was driven by man-power. The man inside the wheel moves the gears, and the movement is transmitted to the part below which sucked up the water to the channels at the top.

Another highly appreciated engineer of the time was Vittorio Zonca (1568–1602), whose work, "Novo teatro di machine et edificii per uarie et sicure operationi" contains several hydraulic machine designs, like the one in Fig. 5.16.

Figure 5.16 shows how horse-propulsion is used to turn the gears that move the wheel that picks up the water which is then poured into a channel where it can be easily collected. Instead of using a wheel with containers that are attached to its external circle to collect the water, a wheel with internal pipes is used so that the water can be poured into an output channel. It can be noted that the current flows in the opposite direction to the wheel motion when collecting water. This is why a horse is required to provide the machine's driving force.

With the aid of his knowledge of hydraulics, Georg Böckler (1648–1685) ("Theatrum Machinarum Novum", 1661) designed new systems for raising water, including fountains and even machines that were thought to be for perpetual motion thanks to the recirculation of water.

There is a period of almost 200 years between Di Giorgio's illustration and the designs by Böckler. The evolution in the quality of drawings, as well as mechanism designs, can be considered to be relevant.

Fig. 5.15 Ramelli's suction pump assembly (From "Le Diverse et Artificiose Machine" [93])

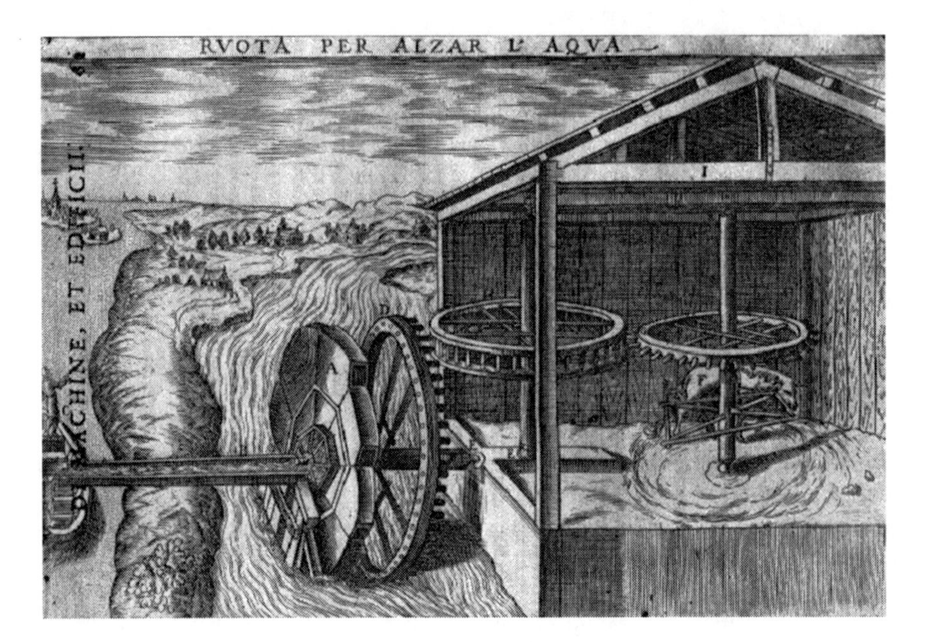

Fig. 5.16 Zonca's mill (From "Novo teatro di machine et edificii")

Figure 5.17 shows one of Böckler's innovative designs for raising water from a well. Animal-power turns the gears that drive a shaft with a properly shaped gear guiding a articulated chain with a series of cylindrical pistons.

The pistons pass through a hollow cylinder that is installed in a well with a radius that is slightly larger than that of the pistons. The piston chain passes inside the cylinder and gradually draws the water upwards to the outside through a tap connected to the hollow cylinder. The main drawback with this machine was the friction resulting from the contact between the cylinder and the pistons.

Jacobus Bessons wrote the "Theatrum Instrumentum et Machinarum", published in 1578, extending the diversity of mechanisms with innovations from earlier machines. Figure 5.18 shows a water pump design that uses wind energy to move a set of gears guiding a mechanism for raising water through some vertical pipes to the surface along horizontal channels. The shape of the sails indicates they were made to turn in one direction, but the mechanism would work equally well in either direction.

On Machine Tools

The Renaissance brought considerable development both to commerce and industry. In order to meet production needs, machines were designed to achieve more efficiency to activities which, up to then, had been considered the work of craftsmen. One example of this machinery is the saw machinery.

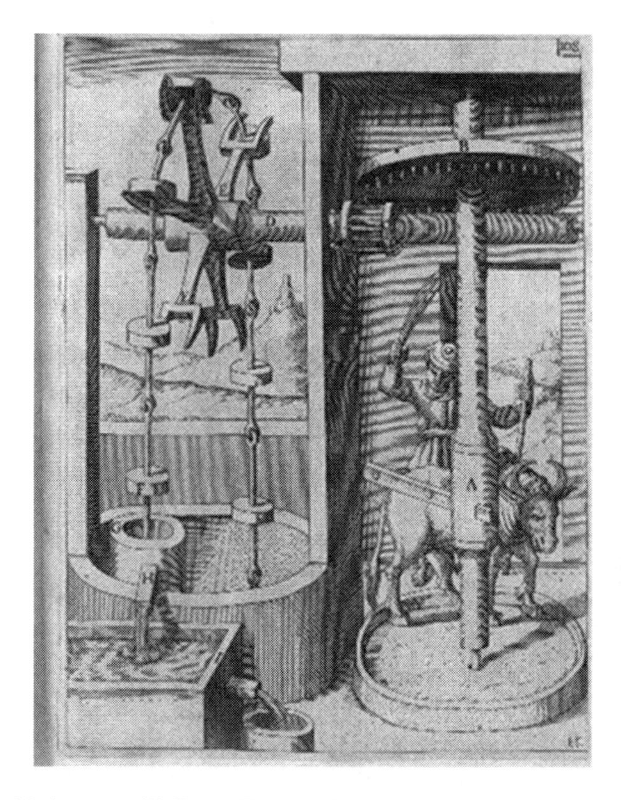

Fig. 5.17 Böckler's water mill (From "Theatrum Machinarum Novum" [24])

In the case of the hydraulic saw, Fig. 5.19 by Di Giorgio is one of the first sketches of this type of machine, after Villard de Honnecourt's sketches in Chapter 4. The plank of wood to be cut is placed on a wheeled carriage, which moves as the saw makes a vertical cutting movement. This movement is provided by a water-wheel that drives a crank and connecting-rod system.

Figure 5.20 is an illustration by Jacobus Bessons, who made a drawing of a saw machine that is powered by a man through arm and foot movements. The force developed by the arms moves a lever and turns two screws with their threads going in opposite directions, as can be noted at the top of the drawing. Then, the screws come together and lower the scissor mechanism along with the saw frame which cuts the wood with its vertical movement. The wood is moved forward by the foot-operated wheel.

Another of Europe's important authors was Jacobus Strada. This Renaissance artist was an art dealer, architect, engineer, and painter. More than anyone else, he exemplifies the absorption of mechanical knowledge in his "Kunstliche Abrís allerhand Wasser – Wind Rosz- und Handt Muhlen" (1617).

An example of Jacobus Strada's work is the hydraulic saw machine in Fig. 5.21 that recalls that of Di Giorgio, but with certain improvements and innovations.

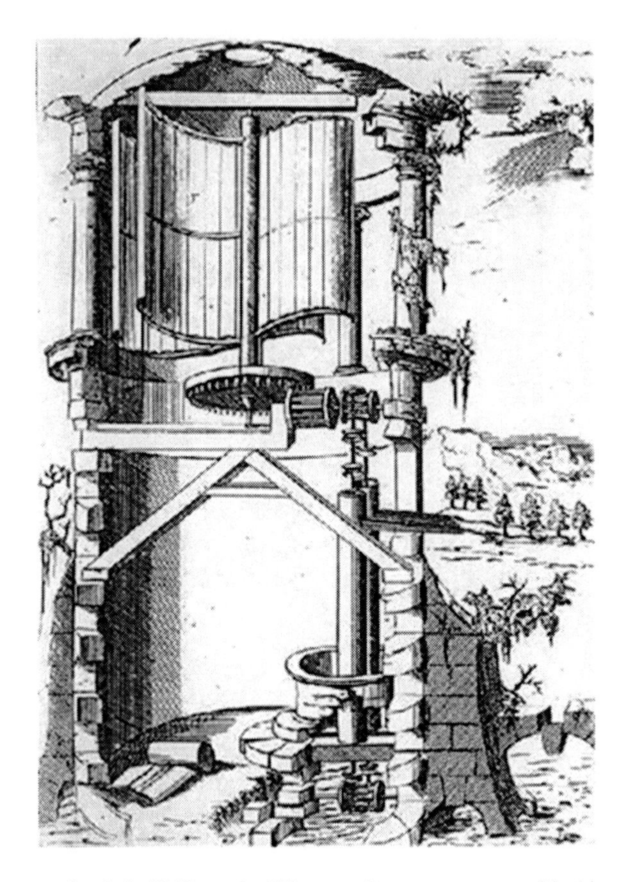

Fig. 5.18 Bessonus's windmill (From the "Theatrum Instrumentorum et Machinarum" [19])

In Strada's new machine, Fig. 5.21, a crank and connecting rod are powered by the flow of water to drive the saw's vertical movement, and there are two pulleys to locate and move the wood to be cut. The drawing is somewhat incomplete as the pulley mechanism is not clearly visible and there is no indication as how the wood is fixed to the carriage.

It should be pointed out that machines and their mechanisms were only partially drawn and their operation was only partially explained in order to defend the inventor's design from copying. Only well-known machines were published while new designs were kept in the "cellar" in the form of technical drawings for a few experts.

On Machines for Traction and Transport

The machinery for lifting and transporting heavy loads, as required by the large building projects of the period, was based mainly on Vitruvius's work.

Fig. 5.19 Di Giorgio's hydraulic saw (From the "Trattato di architectura e machine" [52])

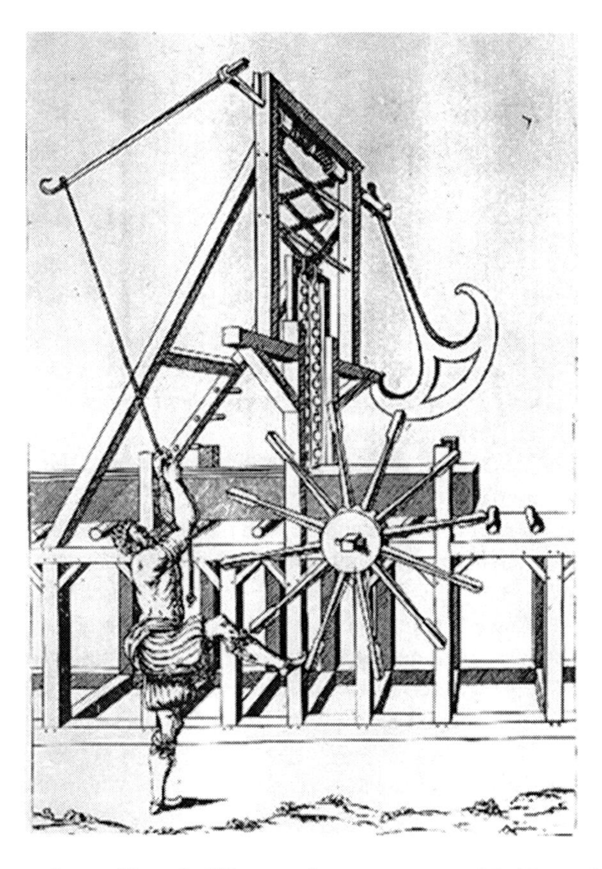

Fig. 5.20 Bessonus's saw (From the "Theatrum Instrumentorum et Machinarum" [19])

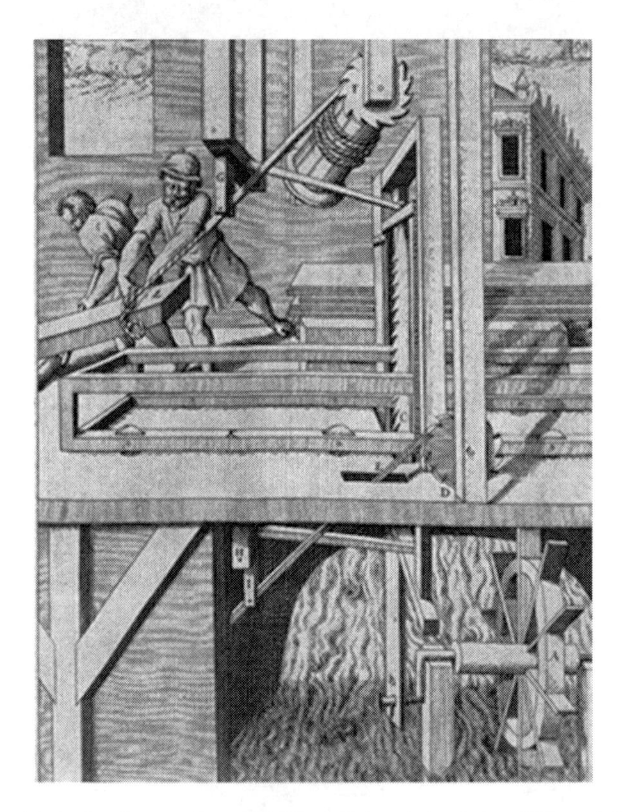

Fig. 5.21 Strada's hydraulic saw (From the "Kunstliche Abrís allerhand Wasser" [117])

However, in the Renaissance, new and original solutions were conceived and developed, particularly for applications in traction and transport. Some examples are illustrated in the following figures.

One of the designs that has aroused much interest is that of the so-called three-wheeled wagon, drawn by Leonardo, shown in Fig. 5.22.

Two large springs supplied the machine with the necessary energy to move. This energy is transmitted to the gearwheels, which are directly connected to the springs. The transmission system delivered the movement to the rear wheels too.

Later, Agostino Ramelli made his drawings with such great clarity that he was the inspiration for a large number of future fields.

The machine in Fig. 5.23 served to pull heavy weights on rollers. The illustration shows a gear train that acts on rope-pulley transmissions to move a heavy set of instruments.

One person is needed to operate a lever and to move the whole machine, but help from other men is required to position the roller cylinders over which the machine will move.

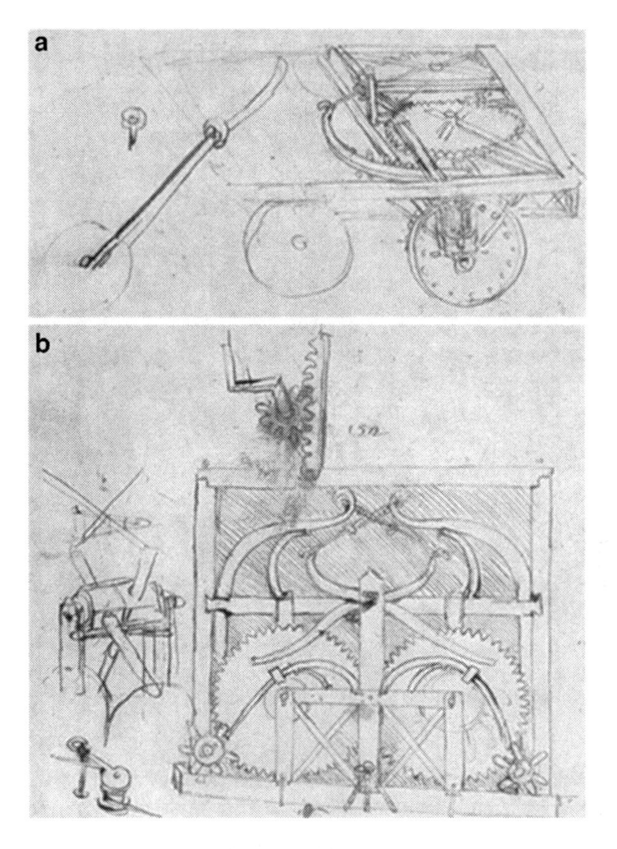

Fig. 5.22 Wagon (From the "Atlantic Codex" [133])

The machine recalls the movement in the Egyptian drawing seen in Chapter 3 (Fig. 3.1). Beside the gear mechanism, the main difference is the use of roller cylinders instead of lubricants to make the load easier to move.

On Machines for the Rural World

The rural world also benefited from the machinery designs to produce agricultural and animal products. New factories were built to develop mass-production in sectors like wine, flour, and wool.

Figure 5.24 shows a model by Zonca that was used to wash woollen cloths. A water wheel propels a drive shaft with some rudimentary but efficient cams that drive some hammers which beat and clean the wool.

Zonca also drew illustrations of wine presses, like the one in Fig. 5.25. It is a large indirect press with considerable capacity. The grapes to be pressed are placed at the base under a large board that is trapped under a large beam. Pressing begins

Fig. 5.23 Ramelli's traction machine (From "Le Diverse et Artificiose Machine" [93])

when the handles are turned to increase the pressure of the board, and the grape juice is collected in a pail. Other presses of this kind are reported in Chapter 3 (Fig. 3.20) although of considerably smaller size.

The machines in Zonca's book are simpler than those by Ramelli or Bessons but they were the current machines that were used at that time for everyday work. Rather than showing signs of improvements, the drawings show efficiency and utility.

Another remarkable machine was made by Jacobus Strada for flour milling as in Fig. 5.26. A horizontal water wheel is used. It should be pointed out that this wheel constitutes an authentic turbine with spoon-like blades that are designed so that the flow will hit the blade and will leave in the precise direction that is required to obtain maximum performance. The wheel drives a shaft on which a lantern is installed to engage with a large horizontal wheel. The transmitted power is divided and shared between the two millwheels that can be seen in the upper part of the illustration.

On Domestic Apparatus

During the Renaissance, machinery also appeared in home environments for reasons of utility and convenience, as is the case for device by Jacobus Strada in Fig. 5.27. The mechanism in the drawing is for fanning diners. The machine is

Fig. 5.24 Zonca's clothes washing machine (From "Novo teatro di machina et edificii")

Fig. 5.25 Zonca's wine making machine (From)

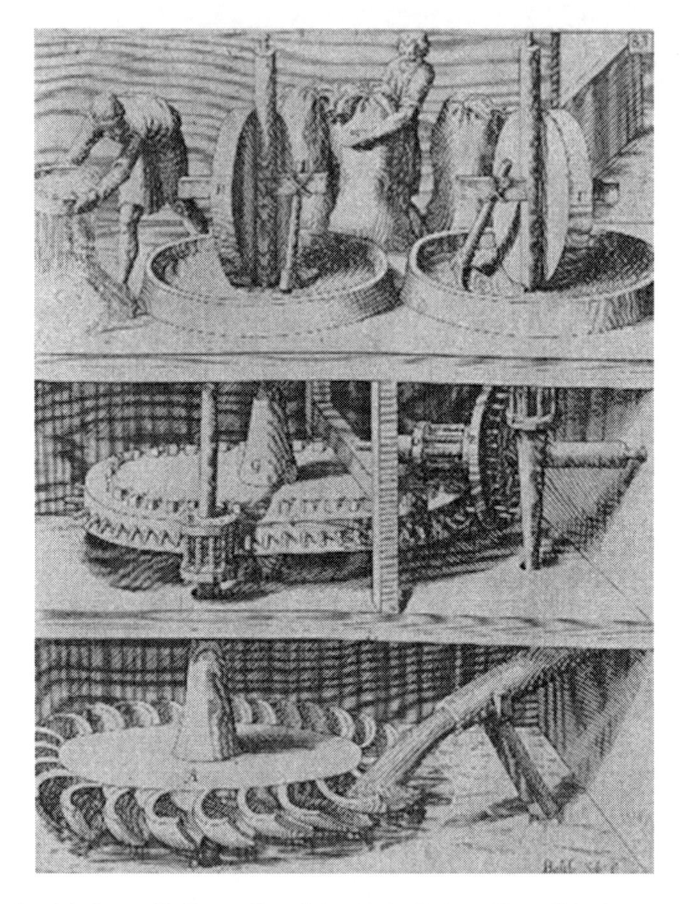

Fig. 5.26 Strada's flour mill (From "Kunstliche Abrís allerhand Wasser" [117])

operated by descending weight that needs to be frequently lifted up to keep the device working. The drawing is self-explanatory, but it should be noted that this automaton was not a decorative or prestigious element like those of Antiquity, but a utilitarian apparatus.

Even private fountains and gardens housed resourceful automatons. In this contest Böckler was a great specialist in hydraulic engineering and his book contains illustrations of fountains, water jets, and designs for irrigating gardens.

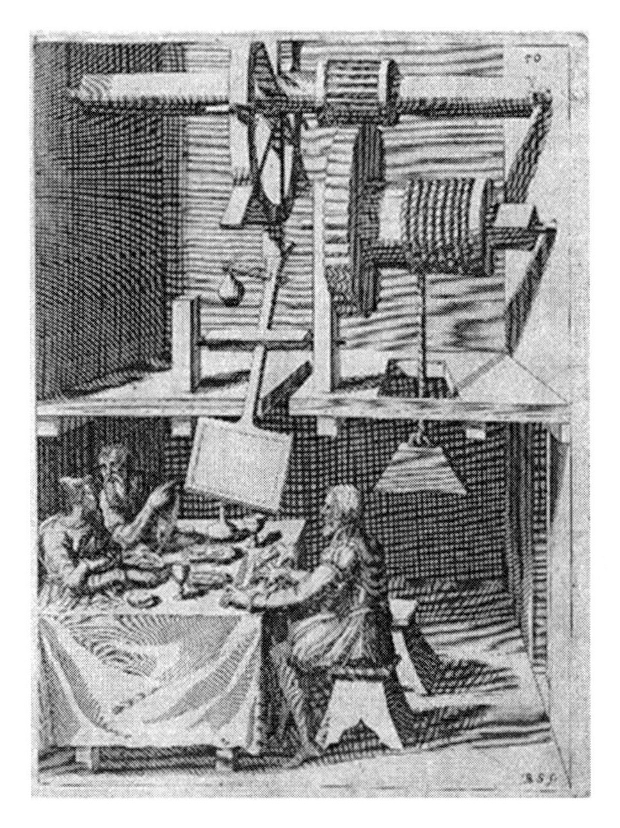

Fig. 5.27 Strada's fanning mechanism (From "Kunstliche Abrís allerhand Wasser" [117])

Chapter 6
Machines in the First Colonial Empires

For different historically well-known reasons, the Modern Age began in Europe and it expanded with its political and cultural influence to other geographical areas by establishing European colonial empires, which have endured to the present day. The influence of European technological superiority is evident in society evolution, and mechanical engineering played a relevant role.

In this evolution, Spain and Portugal had begun to overtake other European nations since the end of the fifteenth century, which is why the first colonial empires were basically from the Iberian kingdoms. Specific centres were founded for the study of science and technology that facilitated this worldwide geographical expansion.

Going round the world was an exploration task requiring mechanical engineering but later, keeping the new trade routes as permanent ways needs a lot of engineering support. Overseas mineral resources required new machinery to mine them.

During the sixteenth and seventeenth centuries, the two Iberian empires were unified for long periods so that the Spanish Crown was the most powerful both from political and economic viewpoints. Thus, the Spanish skill and influence in navigation permitted Spanish establishment in all five continents. It was said that "the Sun never set" over the dominions of King Philip II.

During this Golden Age, the rapid rise of literature, the plastic arts, and music took place simultaneously to the lesser-known but highly relevant developments in other fields like engineering and architecture. This occurred mainly at the end of the sixteenth century and the beginning of the seventeenth century. Hydraulics saw a spectacular evolution with the construction of waterwheels, irrigation ditches, dams, mills, aqueducts, and other water-based devices for the production of power and distribution of water.

Within this framework, a need arose for highly qualified professionals as engineers in the service of the Spanish Crown. Nevertheless, a scarcity of those technical professionals is made quite evident in a letter written by Francés de Álava, in the second half of the sixteenth century in which he wrote: "The persons I know in Spain who are engaged in the service of His Majesty as engineers.... All of them are foreigners, and I do not know a single Spaniard who knows the half of what they do, although I have racked my brains...".

E.B. Paz et al., *A Brief Illustrated History of Machines and Mechanisms*,
History of Mechanism and Machine Science 10, DOI 10.1007/978-90-481-2512-8_6,

The strategic interest in mechanical engineering also remains evident with the milestone publication "The Twenty-One Books of Devices and Machines", which was written around the year 1570 by order of the Catholic Philip II, King of Spain. This work are a large number of machine designs from the period that are classified according to their functions. There are descriptions of pumps, mills, cranes, and other machines, mainly powered by water, wind, gravity, or animal traction.

This machine encyclopaedia, as some historians call it, is of unknown authorship. It was initially attributed to Juanelo Turriano who was famous at the time for the invention of a water-raising device in Toledo. However, later studies seem to credit the work to Pedro Juan de Lastanosa or a combination of different authors.

Another outstanding personality was Jerónimo de Ayanz y Beaumont, who in 1606 was granted rights for inventions (something like patents) for more than 50 devices. This happened after the most prestigious scientists in the kingdom, at the request of King Philip II and his successor, were convinced of the utility, the correct working, and the rigorous scientific methods on which all those machines were based.

The technology rapidly spread through manuscripts, which were frequently the work of unknown authors. An emblematic example of this is a single volume written by Francisco Lobato, an inhabitant of Medina del Campo who, in less than 40 pages, compiled notes on technology between 1547 and 1585.

These popular reflections were little appreciated at the time, which is probably why most anonymous reflections have been lost. This would have been also the case with Francisco Lobato if he would not have written down his reflections on the back of a copy of Ptolemy's geography (published in Rome in 1508).

On Raising Water

Pumping water was fundamental to mining, as well as for supplying water to cities and agricultural environments. Water also had to be baled out of ships during long voyages.

Among the notes that Francisco Lobato wrote "so as to remember" and "so his children might know", is a description of a machine for raising water from a well to a height of two persons, as shown in Fig. 6.1. It was based on what the author calls a screw device, which is an Archimedes screw that is driven by a gearwheel.

Its peculiarity is that the author considers that one-third of the water raised can be used to turn the gearwheel and keep the device going, while the other two-thirds can be used for irrigation, driving a mill.

In the work "The Twenty-One Books of Devices and Machines", several water-raising devices, like Archimedes' screw (shown in Chapter 3) were mentioned and illustrated. They include a machine that is based on several spiral screws of this type in series to thereby enhance its water-raising capacity. It also makes extensive reference to other hydraulic devices such as pumps and water-raising mechanisms.

Figure 6.2 shows a force pump similar to that of Ctesibius, as shown in Chapter 3. The text recommends using these pumps to raise water to great heights, but only in small quantities due to the weight of the water.

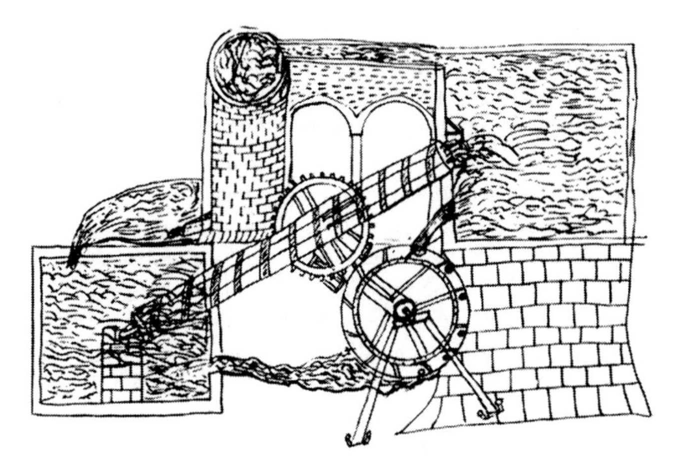

Fig. 6.1 Machine for raising water (From a well from Francisco Lobato's manuscript [50])

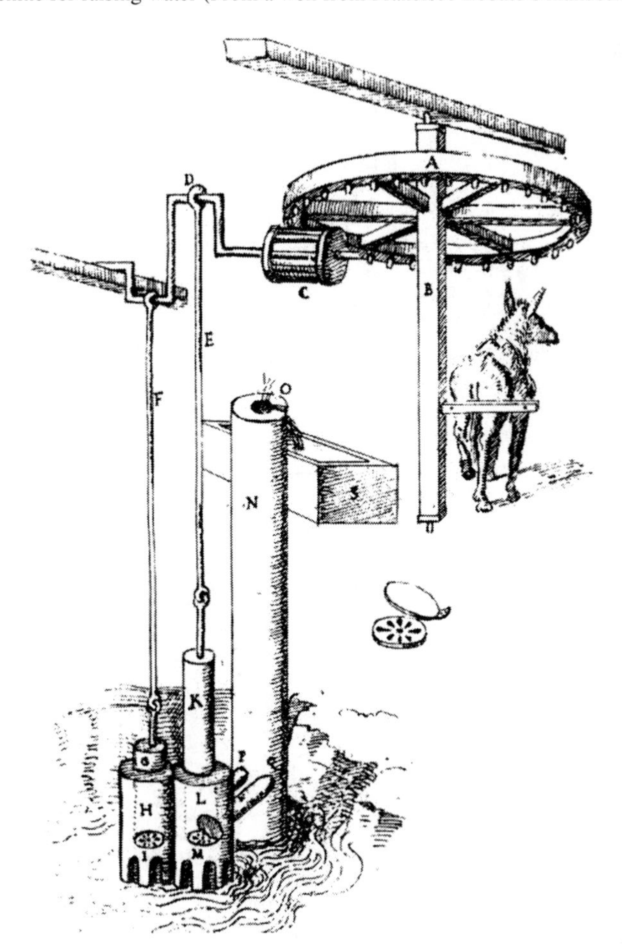

Fig. 6.2 Force pump (From the Twenty-One Books of Devices and Machines [3])

The text underlines the importance of using machine parts of the right size. The traction is animal, which is indicative of the machine's power, and it shows a worm-gear assembly as well as a crankshaft that drives the pump alternately by forcing a flow from one of the two compartments into which the pump is divided. The water passes through a short, horizontal pipe to a tube that pours it into a tank

Jerónimo de Ayanz invented a screw pump, Fig. 6.3a, that is also based on Ctesibius's design and on other similar devices that were used during the Renaissance. It differs for the cylinders that are threaded inside and are operated by a gearwheel to drive two lanterns mounted on the screw axle. The pistons are positioned at the ends of the screw, and they screw into the cylinder, producing high pressure in a flow of water. It worked smoothly without the abrupt movements of earlier piston pumps.

He also invented the pump in Fig. 6.3b, which was highly efficient in pumping water out of ships in the event of a storm or a gunfire attack. Two persons were sufficient to operate the four pistons by using a simplę rocker-arm with counter-weights. Transmission is made by two lanterns, which simultaneously engage two racks that are attached to the rods. The end of each rod has a piston so that, at any

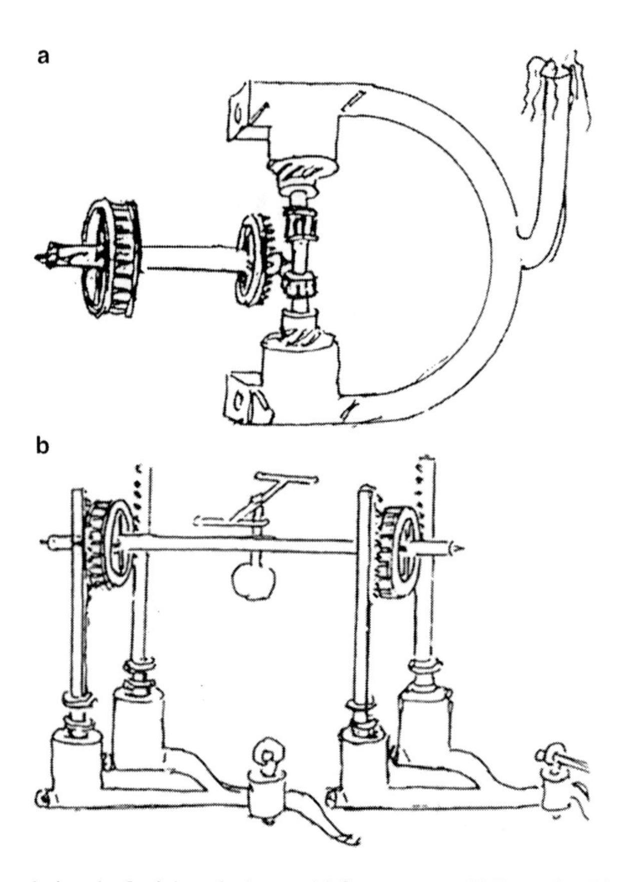

Fig. 6.3 Pump designs by Jerónimo de Ayanz: (**a**) Screw pump. (**b**) Pump for ships [49]

time, two pumps are at the intake phase and the other two are at the outlet phase. In and out pipes are connected to each of the four cylinders. Discharge valves can also be noted in the illustration.

The most ingenious system of the time for raising water was prepared around 1565 to mitigate the need for water in Toledo as a result of its demographic growth. Juanelo Turriano built a device capable of overcoming the more than 100 m difference in the level between the river Tagus and the reservoirs of the Alcázar in Toledo. With a flow of over 12,400 L of water per day, the task would seem an impossible challenge if we bear in mind the apparatus available during that period.

The current of the Tagus itself served as the driving force as well as supplying the water needed for the city. It was the machine's size that made it significant as a work of mechanical engineering, since its constituent parts had already been described in Arab treatises, as reported in Chapter 4.

We have no complete drawing of Juanelo's device but there is a manuscript description with a rough diagram that is shown in Fig. 6.4, as made by the Precentor of Évora (Portugal) in 1604.

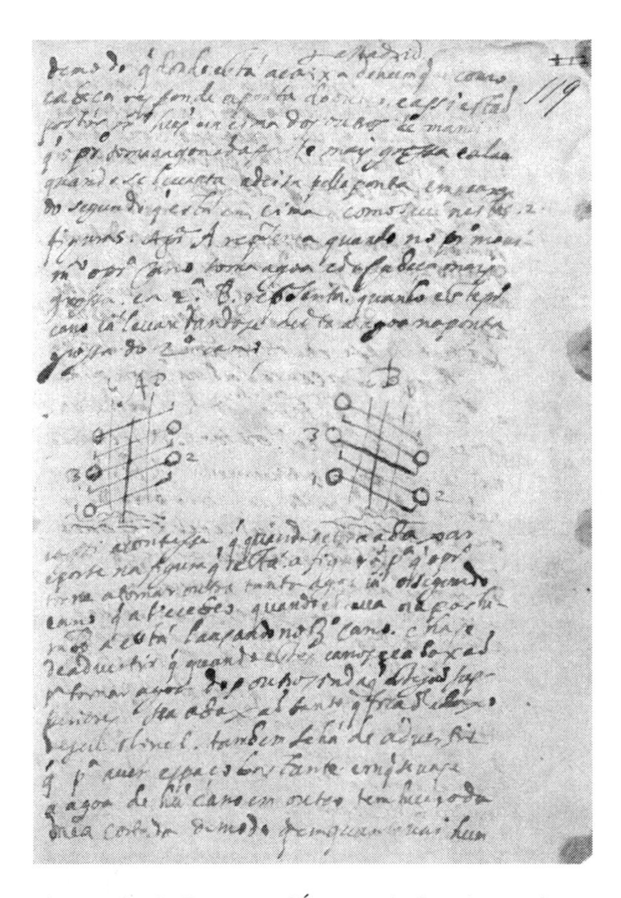

Fig. 6.4 Manuscript page by the Precentor of Évora on the Juanelo pumping system [2]

Both the device and way it worked were very curious at that time. These facts contributed to its inventor's fame and the name by which it became popularly known, "The Dancing Machine". A contemporary traveller named Kenelm Digby wrote, when referring to the device, "...and so the two sides of the machine were like two legs that trod the water in turn". This gives strength to a reconstruction that has been proposed by N. García Tapia concerning the existence of arches for collecting water from both sides of a vertical system.

Juanelo made use of previous knowledge of water-raising procedures. His machine did not represent any actual progress in mechanical systems, but it was undoubtedly a great feat of engineering for the time, both in size and complexity. The size of its parts and the involved forces were a considerable challenge for the technical know-how of that period. In addition, since it was a machine with so many moving parts, it is to be noted that the dynamic effects would be significant, and its joints were subjected to considerable tribological actions. From today's viewpoint, if it was not for the eyewitness accounts that have reached us, it could be doubted that the machine really worked other than as a model or as a curious mechanism to amaze visitors, but not as a supply system for a town with a large population. Juanelo evidently avoided suction systems, not only because of the height but also because of the amount of flow to be pumped, and therefore he based his machine on a series of cups that were installed on arms as the most appropriate solution. The different documents of the period and later studies enabled Ladislao Reti to make a fairly close reconstruction of the mechanical device, in 1967, which is shown in Fig. 6.5.

The illustration shows how the river current operated two waterwheels with paddles. The first waterwheel moved a mechanism with a waterwheel that raised the water several metres to a reservoir tank. The second used another connecting rod-crank-based mechanism to operate the rocking movement of the vertical systems that are called towers, which enabled the water to be raised from the reservoir under atmospheric pressure.

The machine was designed to overcome any difference in level, since, as the water came in contact with the air, no excesses of pressure occurred due to the pumping height, as would happen in a pipe system. A larger difference in level only required an increase of the number of sequentially connected towers. The tower diagram in Fig. 6.6a shows two working positions for the device that are based on a set of pivoted buckets and cups that raised the water in stages, due to the backward and forward motion of the cups. When the water reached the top, it was fed through some pipes to the next tower, and so on until it reached its final destination.

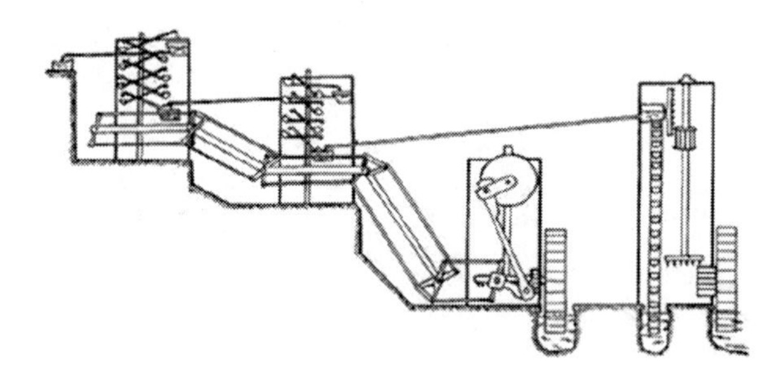

Fig. 6.5 Diagram of Juanelo's device in Toledo. Ladislao Reti's reconstruction

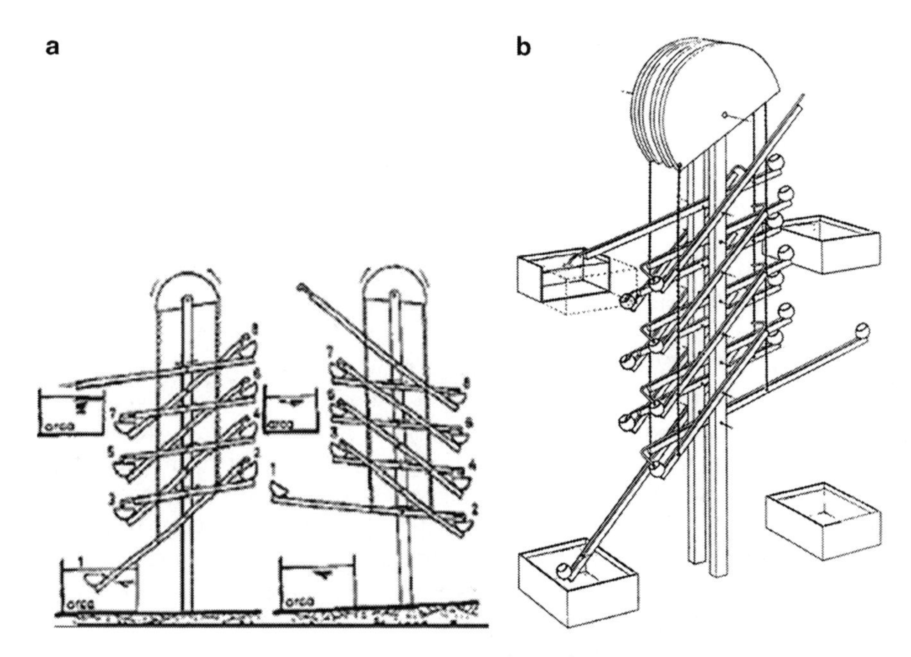

Fig. 6.6 (**a**) Detail of the positions of the towers for raising water using pivoted arms and cups [128]. (**b**) A three-dimensional reconstruction of the tower by Nicolás García Tapia [51]

García Tapia's reconstruction in Fig. 6.6b of Juan Ramos's diagram shows there were two input and two output arches, that were situated either side of the tower. When the first cup was introduced into the arch to take water, the opposite one was raised by passing the water on to the next, and each container was full when its opposite was empty. Thus, the water flowed continuously with a co-ordinated and precise movement.

The "dancing machine" supplied the city during the decades when it was the imperial capital. Former technology advances made such a costly supply system unnecessary, which is why it is now only known through graphic reconstructions and models.

On Mills

Milling was of great economic importance in the first colonial empires because several industrial and agricultural sectors were dependent of it. Many mills were set up under State licence and not only for tax reasons.

Wheat and oil were the staple diet of the founding settlers of the Iberian empires. The words aceña (watermill) and almazara (oil mill), of Arab origin were used in Spanish to denote these machines which gives some idea of how extensively they were used in agriculture. The creation of sugar cane plantations overseas also required the installation of milling machines, to which were significantly given the name of "ingenio", which in Spanish means "device" or "apparatus" with great ingenuity. Mills were also linked to colonial mining.

As for the energy needed to drive them, river courses with sufficient flow were not always at the hand. Wind energy or flesh-and-blood traction (men or animals) were the resources that were used also extensively.

The descriptions found in the Twenty-One Books deal with some simple mechanisms like the "modern flour mill" in Fig. 6.7, that is powered by a flow of water directed towards a horizontal-drive waterwheel. This mill is compared to others, and it is stated that the one in this illustration gives better results than man-powered mills using a rod to turn a crank. In addition, there is a section devoted to the study of the most appropriate slope for the pipes bearing the water flow to the horizontal wheel in order to improve the efficiency of horizontal wheel drives.

Some other illustrations from these books show examples of more complicated mill mechanisms than the above one. Figure 6.8 shows a sugar mill mechanism powered by a waterwheel that combines a speed reducer with a transmission between the lantern and the worm wheel. It is a unique solution since the drive-wheel itself is simultaneously a crown that drives the rod, and a crank mechanism that allows the sugar cane to be cut automatically, ready for crushing. The set of mechanisms enables the cutting and crushing operations in synchronised work.

Indeed, in the sixteenth century, efforts were made to find new sources of energy and work to replace animal power. Not only water and wind energy were used but also the force of gravity for mills, waterwheels, cranes, and pumps were attempted. Along these lines, attempts were made to extrapolate the use of the small, pre-existing mills for designing large cereal mills.

A relevant example is shown in Fig. 6.9 for mill flour where the milling stone is located at the top of the machine on the base that is sketched. The text gives a warning that this type of machine works with counterweights by using heavy weights that are situated at considerable heights. As a result, it had a high milling capacity but it may often have suffered damage to its components. This mill has a more complicated drive mechanism: there are three parallel horizontal shafts, with a

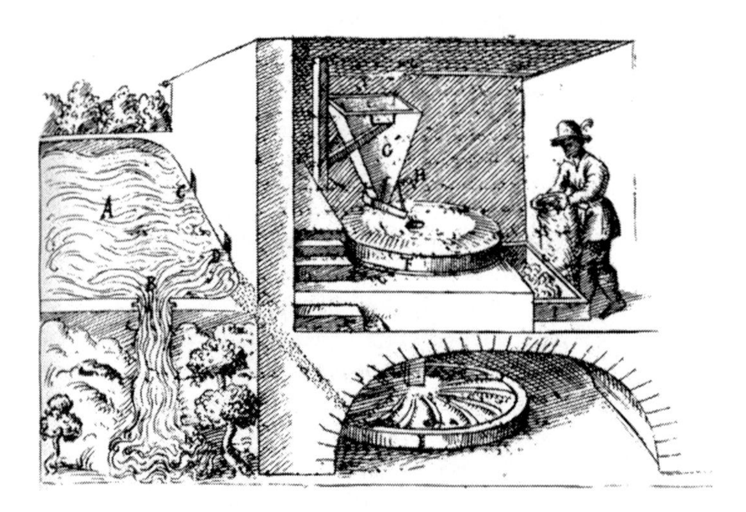

Fig. 6.7 "Modern mill" for flour (From the Twenty-One Books of Devices and Machines [3])

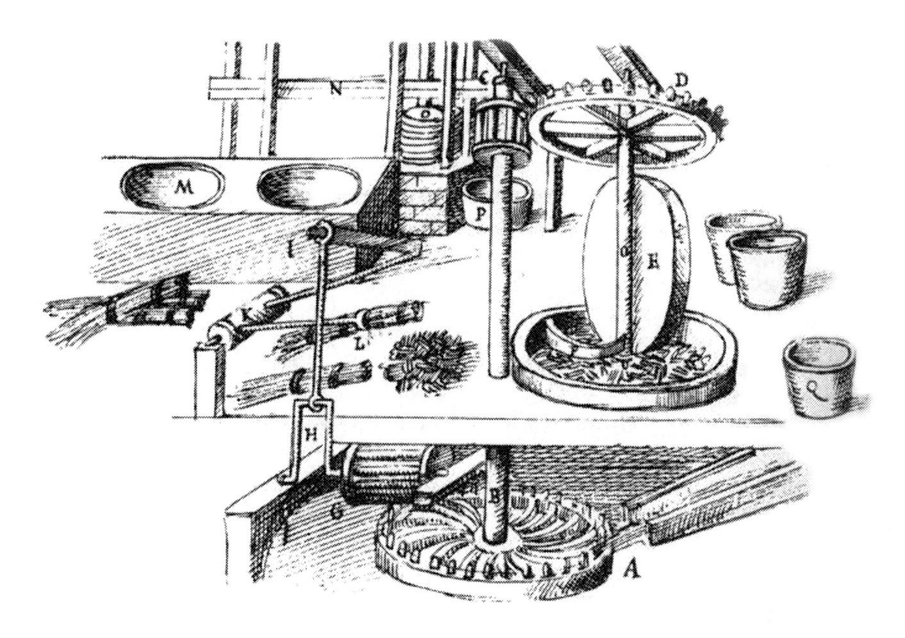

Fig. 6.8 Sugar mill (From the Twenty-One Books of Devices and Machines [3])

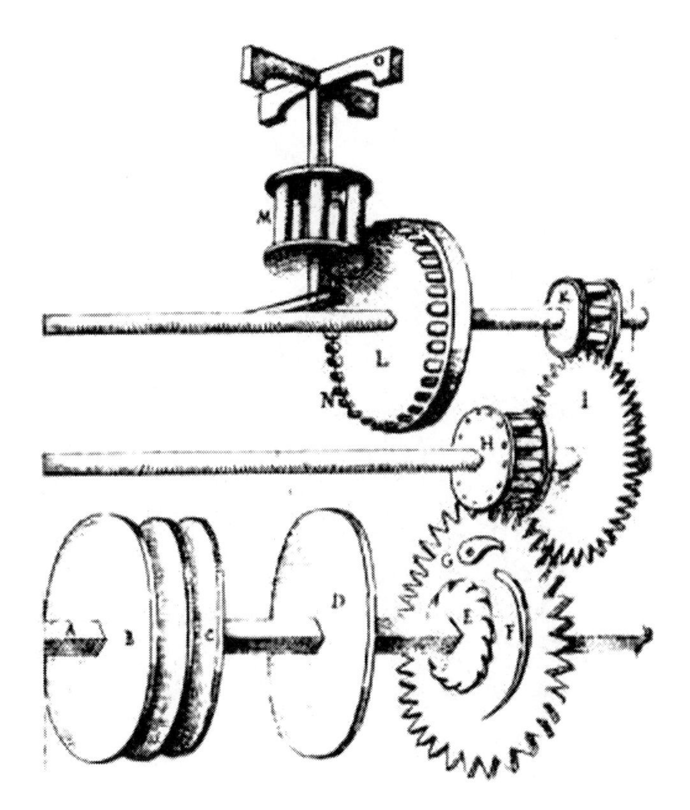

Fig. 6.9 High efficiency flour mill mechanism (From the Twenty-One Books of Devices and Machines [3])

two-stage reduction and an upright outlet shaft. It has two lantern transmissions and a gear wheel that is equipped with a greater number of more point shaped teeth than the wheels in Fig. 6.8. Finally, there is a crown-and-lantern transmission. The input shaft incorporates some additional wheels to serve as a flywheel to get more uniform movement. A ratchet mechanism can also be noted on one of the wheels.

The above counterweight mill design gave rise to lengthy legal litigation concerning the licences for inventions between the Spaniard P. Juan de Lastanosa and the New World settler Ruy Lope de Luna. This legal dispute was settled with a Solomon judgement by dividing the profits between both the designers.

There are other several references to flour mills, like the one in Fig. 6.10, where power is provided by a vertical paddle wheel that is driven by the current of water in the river at the bottom of the machine. This kind of drive element was used when the river course lacked sufficient gradient but had a large volume of flowing water.

In this example, it was interesting that it is possible to adjust the height of the water-wheel and to adapt the position of the paddles according to the level of the water. This was done by turning two screws. The lantern was very long so that it would engage the crown of the noria no matter what height the noria was set at.

Instead of using the force of the water, in other designs of mills, power was supplied by a man or animal (flesh-and-blood mills). Francisco Lobato's manuscript

Fig. 6.10 Adjustable flour mill (From the Twenty-One Books of Devices and Machines [3])

contains detailed descriptions of some of these mills. Although he was a cultured man, his drawings are of poor quality due to his poor experience in machine designs when compared to other authors.

Figure 6.11 shows a mill that is driven with a wheel with tread bars that a man can stand on. The text gives a rough idea of the part sizes, the number of teeth that are required for the wheels and the number of lantern spindles that are needed to obtain a correct transmission ratio.

Jerónimo de Ayanz presented a mill that he had invented with two parallel stages with a flesh-and-blood driven treadmill, but with animals inside instead of men. Figure 6.12 shows this mill together with the gear and lantern transmission to multiply the speed at the output of the device.

Another example of an animal-driven machine is the sophisticated mill in Fig. 6.13, again from the Twenty-One Books, which can run several tasks in parallel. Of especial interest in this drawing is the variety of operations that are obtained

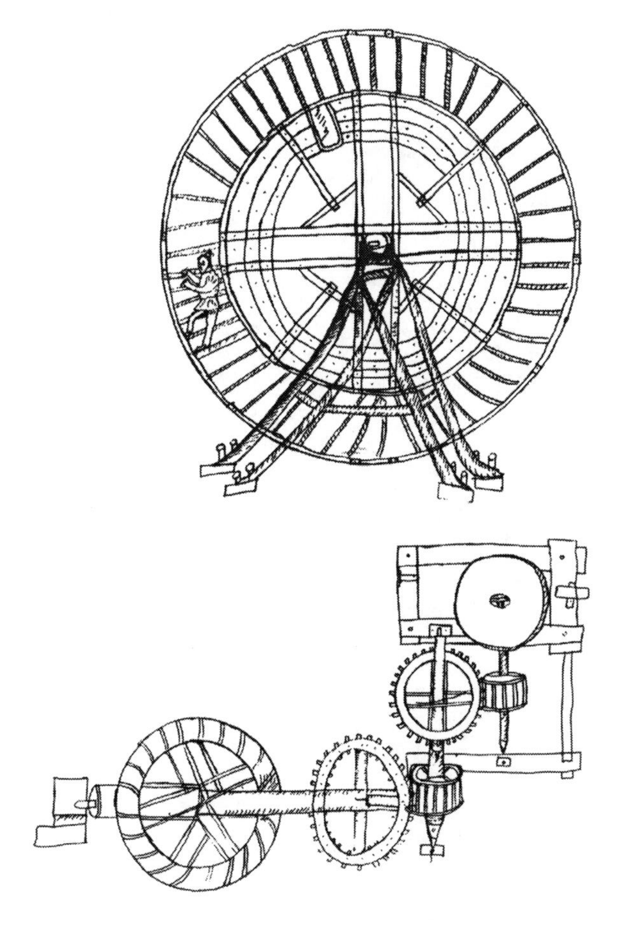

Fig. 6.11 Flesh-and-blood mill powered by a man on the large wheel [50]

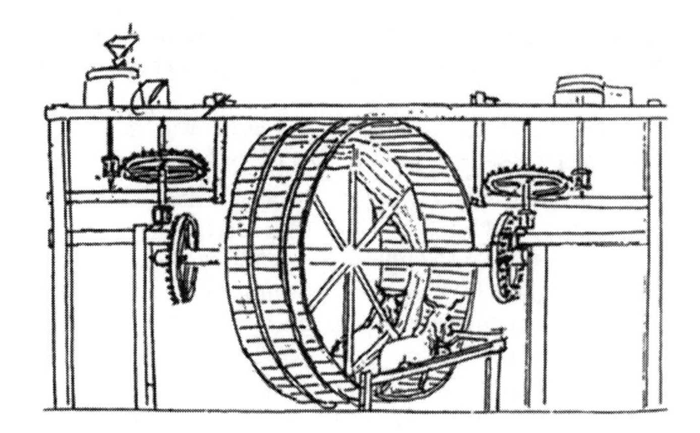

Fig. 6.12 Jerónimo de Ayanz's animal treadmill [49]

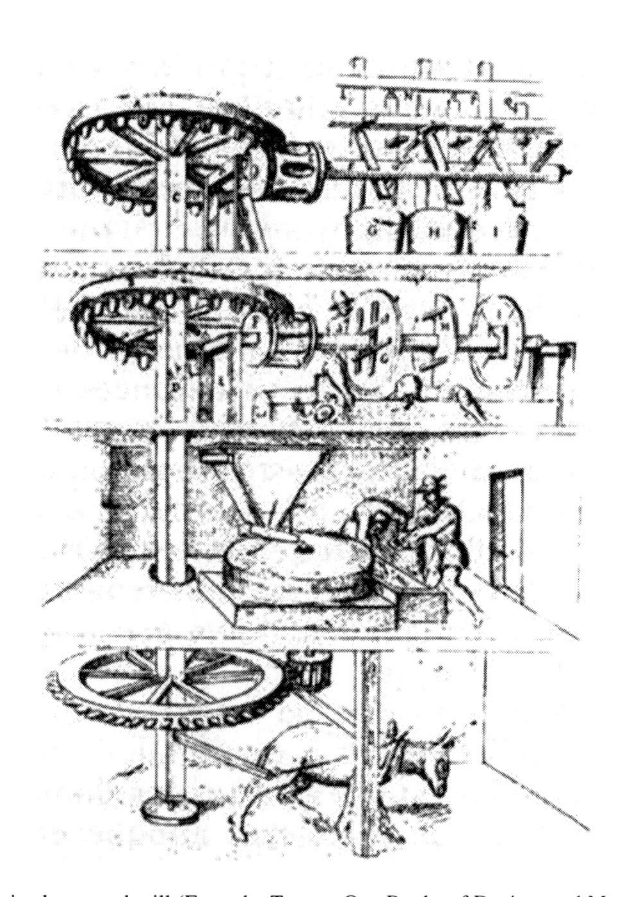

Fig. 6.13 Animal-powered mill (From the Twenty-One Books of Devices and Machines [3])

from the different mechanisms, but all of which are operated by the same upright shaft to which the animal is harnessed.

The machine at the top is crushing gunpowder with two pestles for each mortar. One stage below, there are wheels for cleaning and polishing weapons. Below, there is a flour mill whose design is similar to the previous ones. Advice is reported on how to locate the machine on the shaft support and to separate the housings.

The Twenty-One Books of Devices and Machines also refer to the use of this type of mills in fortresses. One of this machines is mentioned operating in the city of Seville as an exception when it would seem more logical to use the flow of a river. One of the machine stages is used for crushing gunpowder and a series of mortars can be noted at the top of the illustration, as they are driven by a horizontal shaft that is fitted with projecting tongues to act as simple cams.

A discussion is reported on the utility of these mills when they are compared to alternative water mills. The author also gives his opinion concerning the feasibility of a device, which lets us to think that the machine in question never existed but it is rather a design by the author by grouping together other known machines. The animal drive around the upright shaft gives an idea of the machine's size and power.

For other applications, the mill is a windmill, as in Fig. 6.14. Referring to it, the writer assures us that this windmill is very common in Flanders, Germany, and France, but not in Spain or Italy because the winds in these regions are not suitable for driving them as there are no normal and very intense winds in those countries.

Francisco Lobato's manuscript also contains references to the no-longer existing windmill in Almagro that is reported in Fig. 6.15. He claims that this windmill was

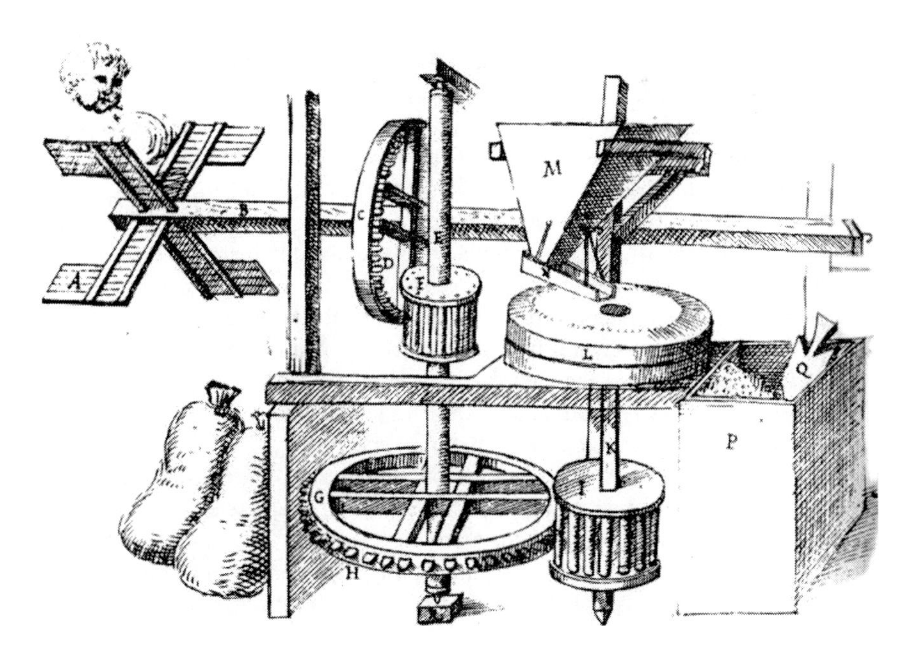

Fig. 6.14 Windmill (From the Twenty-One Books of Devices and Machines [3])

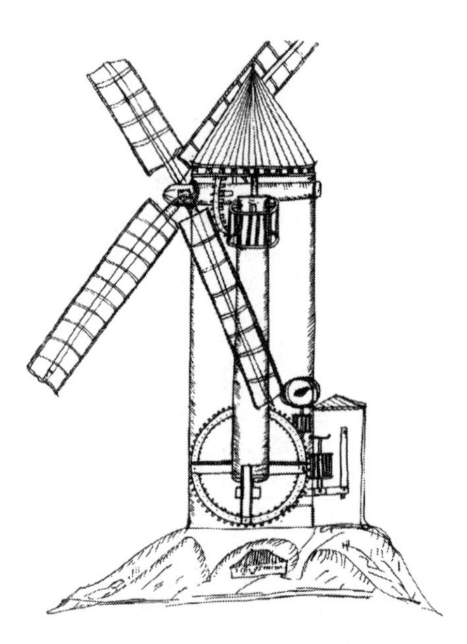

Fig. 6.15 Almagro windmill (From Francisco Lobato's manuscript [50])

the most costly seen in Spain at that time. It was driven by a horizontal shaft that is positioned on a small hill of 40 ft high and it operates several speed multipliers, which is why, the text states "... although the sails turned slowly, it brought great swiftness, alacrity and fury when cutting...".

In many places and particularly overseas, there were fast-flowing but inappropriate rivers for hydraulic power because of their small (or almost non-existent) difference in water level. However, mechanical genius of the time managed to overcome this difficulty by building floating mills.

Figure 6.16 depicts floating flour mills that were noted in many parts of Italy by Jerónimo de Ayanz and the anonymous author of the Twenty-One Books. The anchor allowed them to adapt to the rise and fall of the river. The wheels that are powered by water from beneath had to be placed in pairs or be wider than other types of mill wheels, since the slow running water often did not provide enough force to turn them. Consequently, it was quite usual for them not to be able to grind much grain unless they were on fast-flowing rivers.

Figure 6.17 shows a device for sieving flour with the aid of cranks. The assembly at the right of the illustration is shown as an automated alternative, with a flywheel and two cranks. It points out that the two cranks should be the same size. This machine was important because it was attempting to automate and to increase the production of other processes inside the factories which from the mill houses were converted. The flour drops into box T while the bran piles up at S in an almost non-stop process.

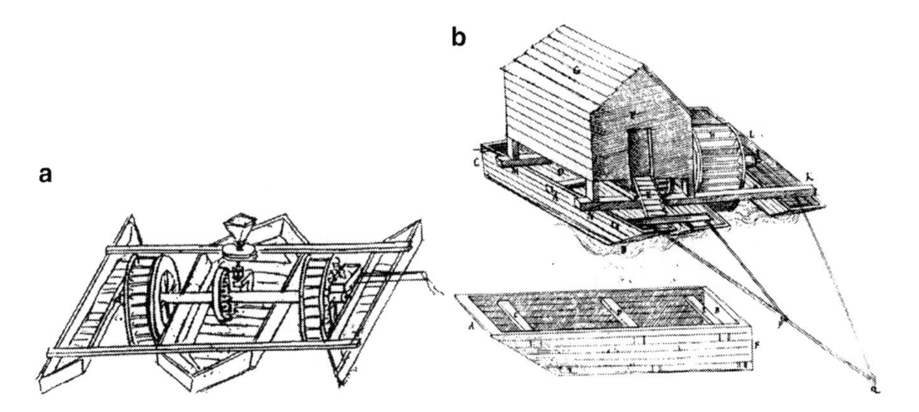

Fig. 6.16 Floating mills. (**a**) By Jerónimo de Ayanz [49]. (**b**) From the Twenty-One Books of Devices and Machines [3]

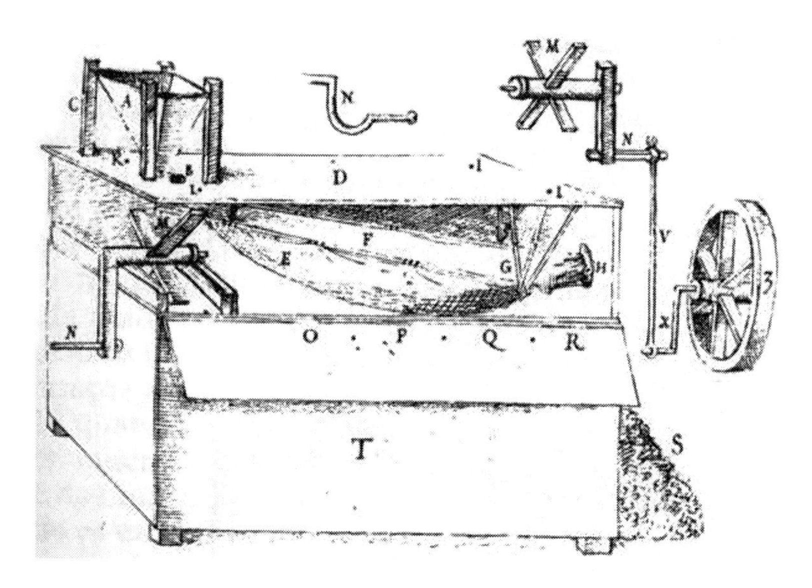

Fig. 6.17 Flour sieve (From the Twenty-One Books of Devices and Machines [3])

Jerónimo de Ayanz's innovative and perfectionist mind led him to replace mill-wheels with bronze or iron cylinders, as Fig. 6.18 shows. Thus, finer flour was obtained in greater quantities, superseding earlier portable milling machines with metal rollers that were designed by other inventors.

At the left of the drawing in Fig. 6.18 there is a lever with stop positions that can be moved backwards and forwards to operate the machine. The work is done by the machine and the operator has only to supervise and intervene without having to make any great physical effort.

Fig. 6.18 Jerónimo de Ayanz's metal roller mill [49]

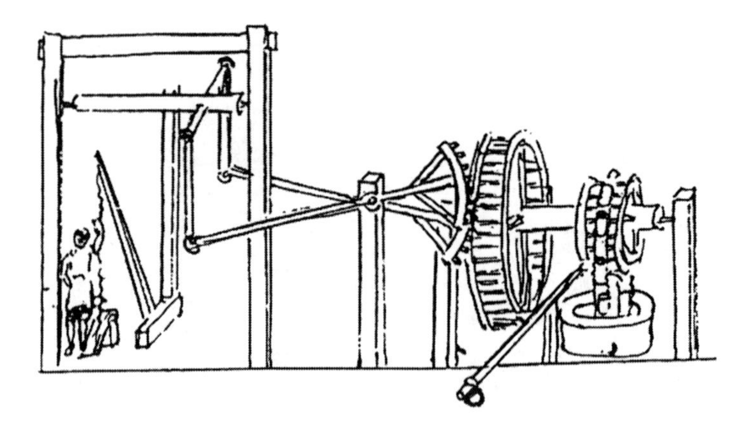

Fig. 6.19 Jerónimo de Ayanz's backward and forward motion machine [49]

On the following page of the manuscript by Ayanz, the more human role of the operator is made more explicit in a civilisation that had declared against slavery. Philip II's moral concerns had led him to organise a social security system for the labourers in El Escorial that included the right to a pension.

Figure 6.19 explains a backward and forward mechanism. A man is able to operate the machine by pressing a foot pedal and pulling a cable with the help of a system of levers and counterweights. The man generates a backward and forward motion on two gear sectors and it is transmitted to a wheel and then to a horizontal shaft whose movement can be used for several applications. In the illustration, the example show how to move a noria with pitchers. This type of backward and forward motion device has some forerunners like, for example, the designs by Leonardo da Vinci.

It is worth noting in Fig. 6.19 the presence of a lever with a counterweight. A comparative measurement of the efficiency of the different machines may be obtained by the position of the weight on the rocker arm.

On Lifting Devices

In the Twenty-One Books there are also devices, like that in Fig. 6.20 that are used for lifting building workers in a kind of cage. Elevation was achieved by some manually operated drums that took up the rope by using a pulley system. It could also be used for bringing down persons like, for example, to mining galleries.

There are comments in the text that point to the advantages of using pulleys and multiple ropes to reduce the weight of each cord when lifting very heavy weights.

The great works of architecture that were carried out during the Iberian Empire also required lifting devices. Many of those devices are attributed to Juan de la Herrera who used them to build the monastery El Escorial. These cranes added nothing new to the designs from the Renaissance period like those that are described in Chapter 5.

On Other Devices

It is not easy to summarise all the other types of machines contained in the "Twenty-One Books of Devices and Machines", which is why only a few examples are given below.

Overseas relations required the enlargement of ports, but mainly the building of new ones. Pile driving machines were frequently used in the period. Figure 6.21a shows one of the machines of this kind. It was manually operated by pulling ropes to raise the hammer with the use of pulleys. The way of joining the ropes to the pulleys and hammer is shown in detail in Fig. 6.21b.

Many of those machines were suited for craftwork in workshops.

The machine in Fig. 6.22a consists of a hydraulic press with a crankshaft to drive an articulated mechanism. This move swing hammers that strike a wedge for pressing by means of a crossbeam. The machine is shown during its operation in Fig. 6.22b.

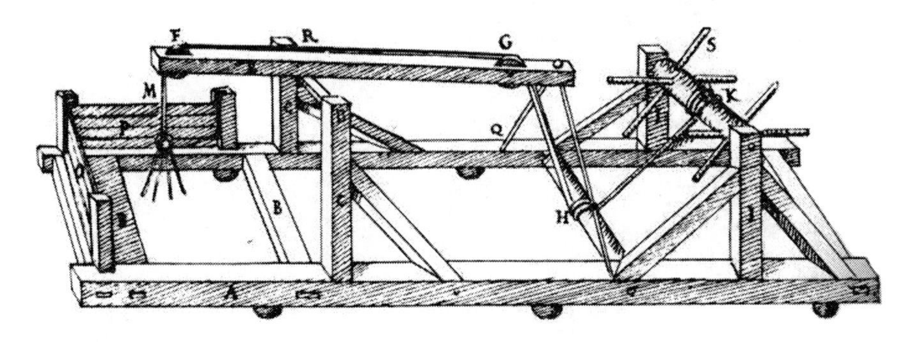

Fig. 6.20 Machine for lifting and lowering workers (From the Twenty-One Books of Devices and Machines [3])

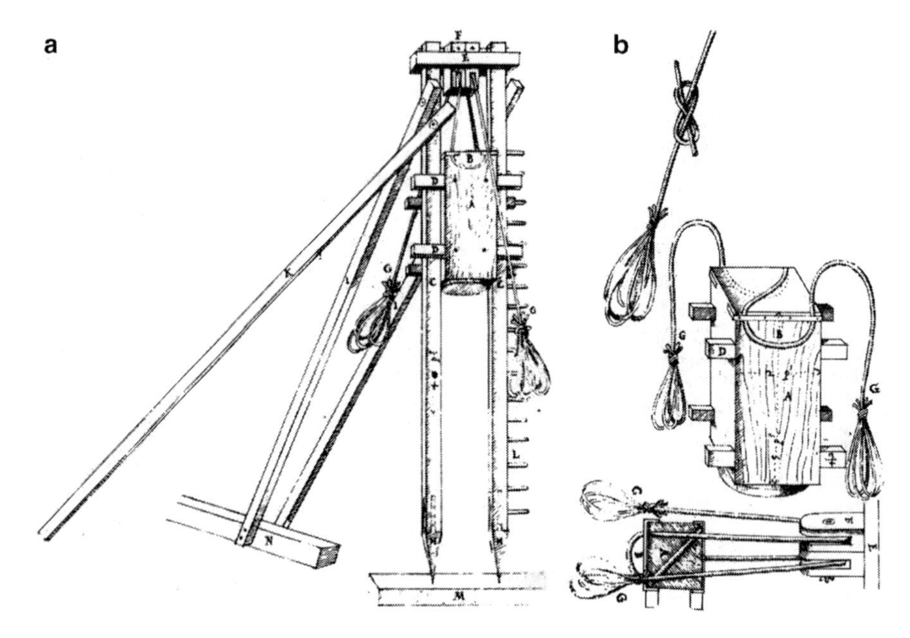

Fig. 6.21 (**a**) Pile driving machine. (**b**) Detail of the device (From the Twenty-One Books of Devices and Machines [3])

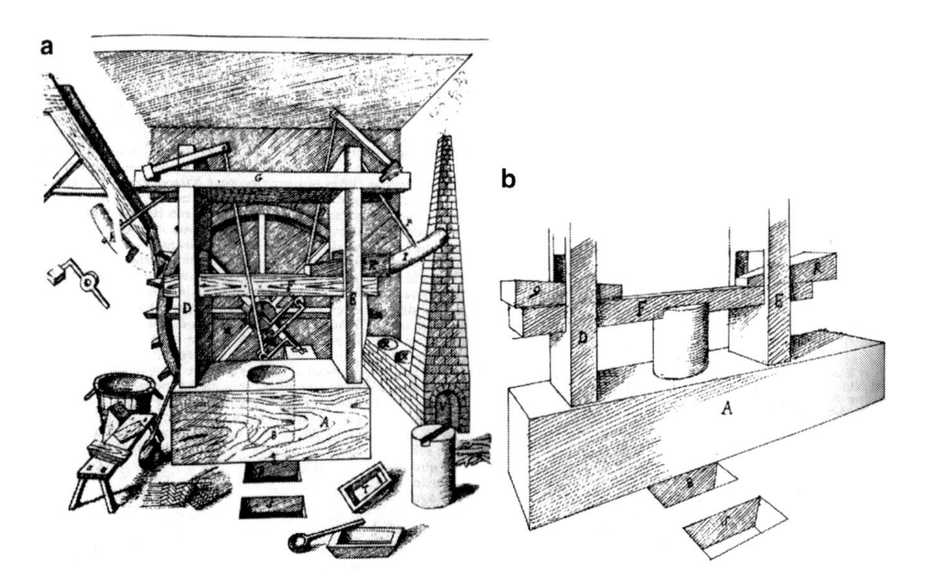

Fig. 6.22 (**a**) Wax press. (**b**) Detail of pressing (From the Twenty-One Books of Devices and Machines [3])

Another device, shown in Fig. 6.23, is as a machine for beating and washing clothes with water. A small flow of water at halfway along the water wheel radius is used to operate the device by producing the movement of a shaft bearing two radial actuators that act as the pendulum motion of the washing components. This is a device that is similar to the one by Vittorio Zonca that is mentioned in Chapter 5 (Fig. 5.24). The difference between the two devices is the propulsion method, which is more efficient in this example because the water supply is constant with a uniform flow, and it will provide an uninterrupted motion.

The shape of the cams is practically the same, which seems to suggest that knowledge was not only transmitted orally but that the author of "The Twenty-One Books" might have seen a drawing of this kind of device (probably Zonca's) or might even have been familiar with a machine that had actually been built.

Fig. 6.23 Clothes washing device (From the Twenty-One Books of Devices and Machines [3])

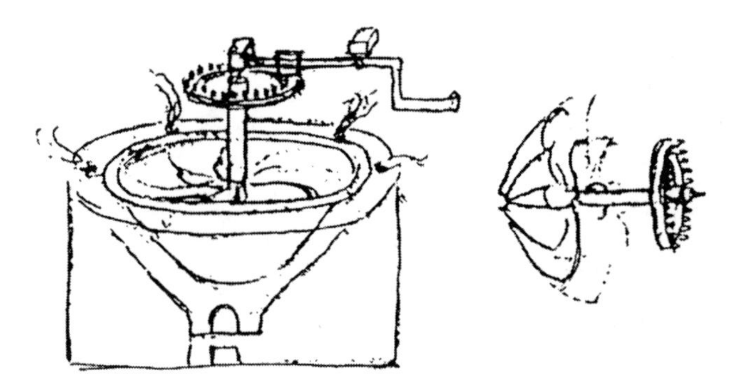

Fig. 6.24 Paddle agitator used in the Potosí mines [49]

On Machinery and Precious Metals

A chapter on the Spanish Empire can hardly fail to mention the gold and silver from the Indies, as this also gave rise to the numerous designs and uses of machinery.

An example of this is the paddle agitator developed by Ayanz in 1606 for processing the silver from Potosí, as shown in Fig. 6.24.

Precious metals were shipped to Europe in ingots, but later it was decided to produce coins directly in the colonies. Coin factories or mints were established for this purpose (like in Mexico in 1535 and Lima in 1565), and they were equipped with suitable machinery. From the time of Antiquity, coins had been made with a die and a hammer, and also with hand-presses, but in the seventeenth century, minting were mechanised.

Specific machines were built for rolling precious metals and making the coins. Minting and rolling machines from Germany were installed in the coin factories on the Peninsula and then in America. Machines were those proposed by private business in 1729 like, for example, for the Lima Mint, as mentioned by Glenn Stephen Murray in his book "The Royal Machine of the Segovia Mint".

In contrast to hand-presses, like in Fig. 6.25a, in 1,725 machines like the one in Fig. 6.25b were installed with lead balls acting as flywheels to accumulate the energy that is required for minting.

The initial process for rolling a metal sheet can be seen in Fig. 6.26, where it is given a prior pass to smooth it. In the book on the Segovia Mint, the writer José María Izaga shows the rolling mill in Fig. 6.27, as installed in Potosí in 1769. The rollers can be noted in the centre of the photograph and they are driven by gearwheels to which movement is transmitted from a lantern. The wheels are made of wood with replaceable teeth, and the entire machine was built in Seville.

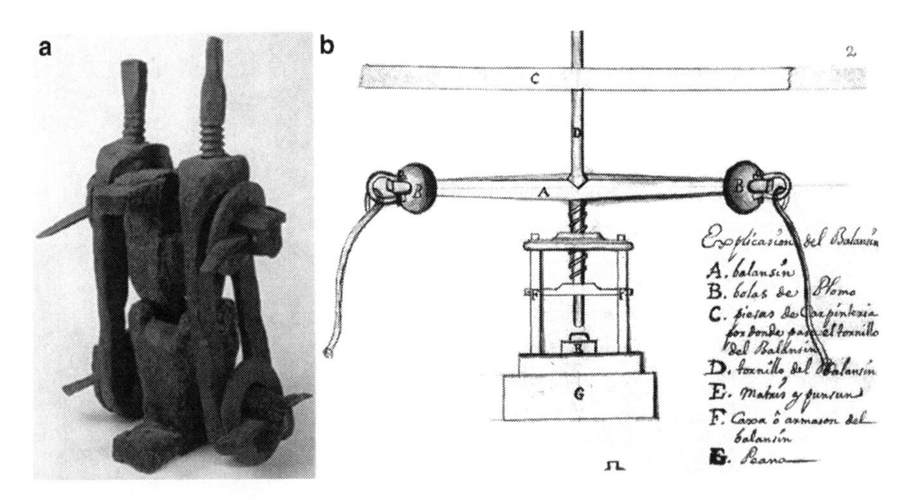

Fig. 6.25 (**a**) Hand-minting press [127]. (**b**) Flywheel minting press from the Lima Mint [85]

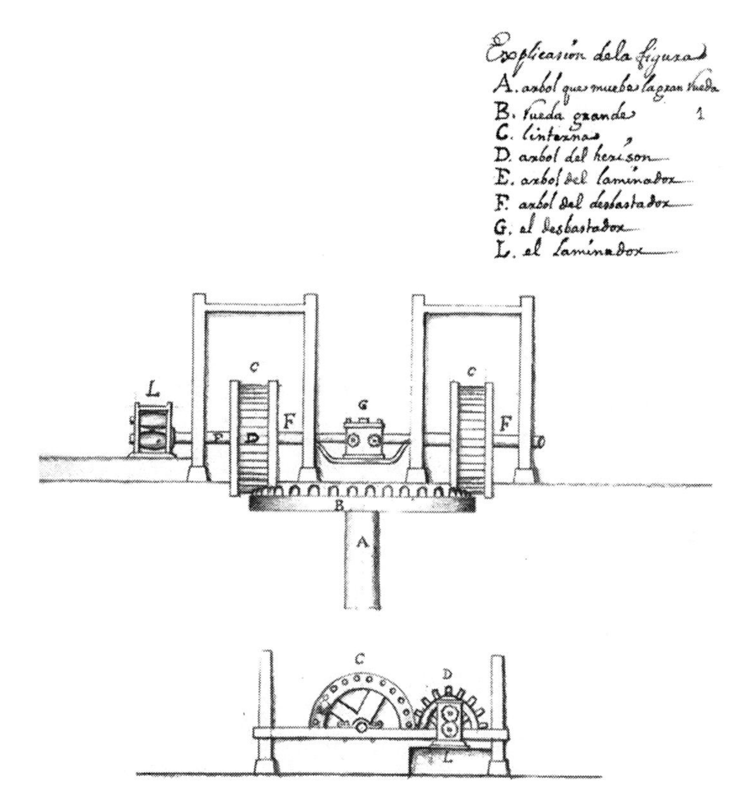

Fig. 6.26 Rolling machine (From the Lima Mint [85])

Fig. 6.27 Photograph of a rolling mill installed in Potosí [85]

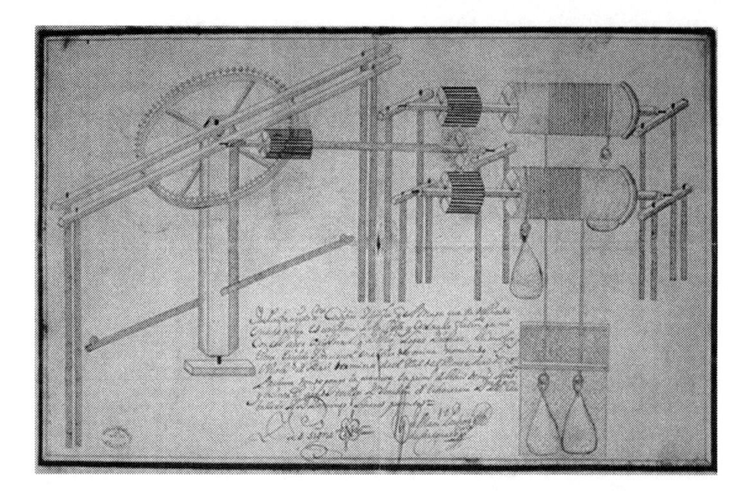

Fig. 6.28 Mine drainage device, used in the San Nicolás de Bari mine (Pachuca) [127]

Other frequently used machines of the time for mining applications were drainage machines, since water was a constant concern in the mines of New Spain. Figure 6.28 shows one of these machines, with no original parts, as it was used in a mine in Pachuca.

On Automatons

In the book by José A. García-Diego on Juanelo's clocks and automatons, the engineer's main activity is made clear that he was a clockmaker for the Emperor as well as an automaton designer. However, there are doubts on his authorship of the most famous works like the automaton of the Lady of Vienna.

Juanelo's popular reputation is a mixture of confirmed historical facts and other stories, but distinguishing the line separating fact from fiction is not easy.

Many suppositions have been made regarding a famous automaton which supposedly accompanied Juanelo through the streets of Toledo. This was known as "The Stick Man", about which written sources and other references exist. Figure 6.29 alludes to the street in Toledo dedicated to "The Stick Man". The text reads: "Toledo Legends, 'The Stick Man'. Along this street passed the wooden automaton built by Juanelo Turriano, clockmaker to Charles V, to the amazement and perplexity of the crowd".

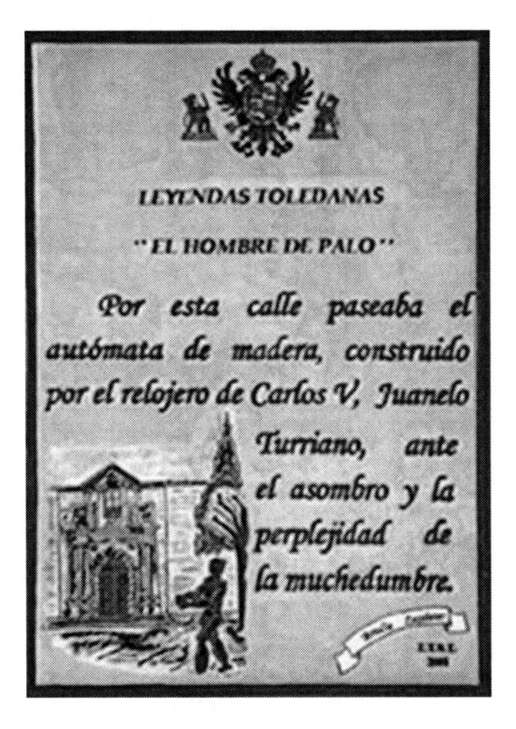

Fig. 6.29 Reference to the "Stick Man" in Toledo

Chapter 7
Machinery During the Industrial Revolution

To attempt to cover machine developments during the Industrial Revolution in just few pages is as absurd as it is impossible. This period of our history arose after the knowledge of previous eras had been accumulated and through a combination of a series of factors that resulted in a period of continuous advances and progress that ended up in a change of approach both in society and engineering, with technical developments of huge quantity.

The construction of the steam engine was a key point, but maybe more at an engineering level as it established a concept: "We don't need men for work, we have a machine". This phrase stated very simply the search for automation in any field, like in agriculture, mining, or the textile industry where machines began to substitute people as a result of the newly discovered technologies.

Thus, there are many people who contributed to this development that not even the largest encyclopedia could describe with so many details. Thus, some of the most significant achievements have been outlined by using beautiful illustrations that are contained in the books on machines of the long fecundus period of the Industrial Revolution.

During the Industrial Revolution, all the fields of technology were improved with a velocity that had never been experienced in the past, although none of them actually had a predominant role, not even the invention of the highly renowned steam engine. These developments may have had a specific influence on and were achieved because of the Theory of Machines and Mechanisms as a result of a society whose needs were increasing day by day.

On Textile Machinery

One of the technologies that experienced greatest changes was the textile industry. In a very short time after 1770, textile industries became divided into those that possessed technology and machinery and those that continued to use manual and antiquated techniques. In order to get an idea of the change, it is worth referring the fact that in 1760 the London "Society of Arts" called for a competition for the best

E.B. Paz et al., *A Brief Illustrated History of Machines and Mechanisms*,
History of Mechanism and Machine Science 10, DOI 10.1007/978-90-481-2512-8_7,

invention of a machine that could be capable of spinning six threads at the same time with the aid of only one person. Technological progress was the name of the game.

In 11 short years, the textile industry was flooded by three machines that brought radical changes to production, namely Arkwright's "Water frame" (or continuous motion machine) that was patented in 1769, Hargreaves' "Spinning Jenny" that was patented just a year later, and Crompton's "Mule" in 1780.

It is probably Arkwright, who most made possible great developments in the textile industry with the machine illustrated in Fig. 7.1. For the first stage of spinning (which has three stages: stretching, twisting, and winding) he invented grooved rollers that continuously stretched the thread.

Hargreaves invented the double carder with the following characteristics: two cards were placed in a normal spinning wheel, one attached to a frame and the other moved by ropes and pulleys so that twice as much work was done. Figure 7.2 shows a further development of this first machine with 12 cards six of which were static and six were mobile. It was not long before "jennies" with up to 100 spindles were built.

Then, Crompton's machine was a result of combining these two machines together. From that moment in textile factories, machines as complex as the 1860 "mule-jenny" in Fig. 7.3 were installed in large numbers. It can be noted how the multiple spindles are located at the left of the machine and how the threads are stretched by means of distance and the grooved cylinders to the right.

After spinning, the next job was weaving. The most outstanding machine for this task was invented in 1801 by Joseph Marie Jacquard. Based on the advances by Bouchon and Vaucanson, he built a machine called the "Jacquard", which worked with the use of perforated cards to weave different patterns of cloth. This perforated card had the task of automatic weaving. Later this idea helped Charles Babbage

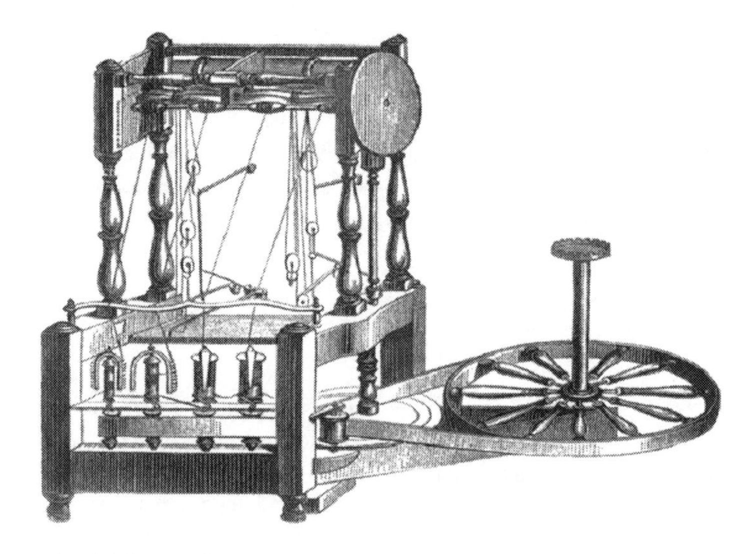

Fig. 7.1 Arkwright's water frame

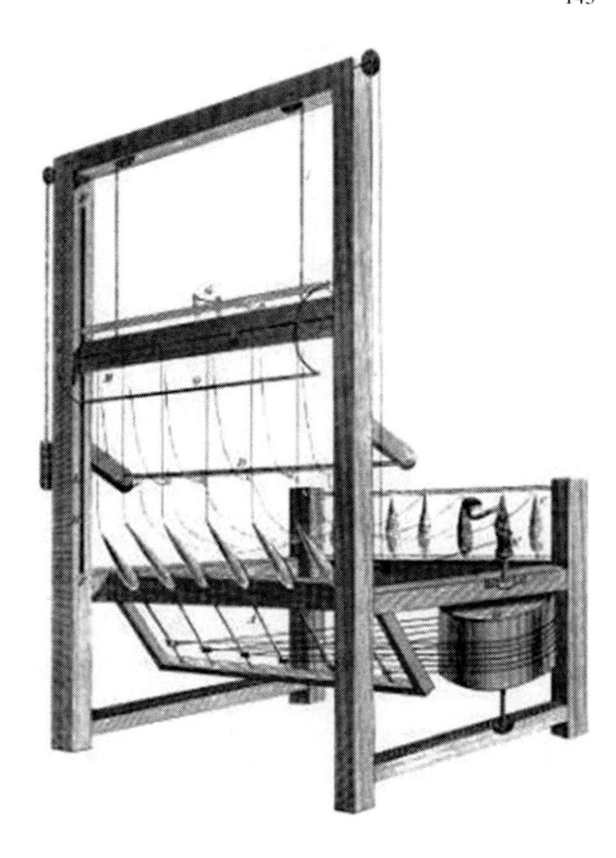

Fig. 7.2 Hargreaves' "Spinning Jenny"

develop his idea of a mechanical universal calculator, which was the precursor of modern computational methods.

Figure 7.4 shows a picture of Jacquard's loom. The man sitting in front of the machine had only to pass the threads and to choose the card for the cloth. Once this operation is performed, it was sufficient to operate the lever that made the threads run and transform them into the cloth that emerged from the top with the required intricacy of threads.

The next stage after weaving was sewing. Bartholomy Thimonnier created the first sewing machine, since he is usually quoted as the inventor, because in 1830 he obtained a patent from the French government to use a latch needle. His machine is shown in Fig. 7.5. It can be noted how the thread is fed to the needle by a sequence of pulleys and springs and it perforates the cloth to produce a chain stitch seam.

One of the sewing machine's main parts is the loop-taker or shuttle. The thread passes through the cloth forming a loop under it. The loop-taker catches it as the needle withdraws and goes on to pierce the cloth again when it has moved forward. The new loop then enters the old one and thus the sewing is completed. One or two threads are used depending on the type of machine.

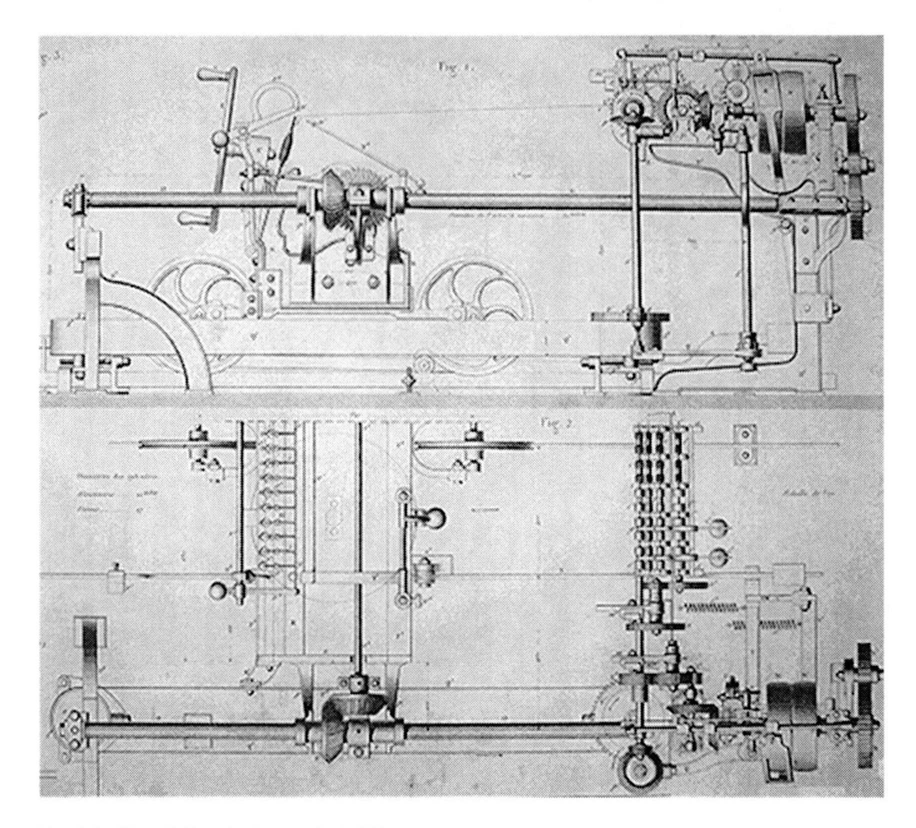

Fig. 7.3 The 1860 spinning mule [124]

Elias Howe's machine was the next new one and appeared in the United States in 1844, and some years after, Isaac Singer (1851) produced an improved version.

Many were the advantages of the new sewing machines: the work became automatic, the stitching was regular and there was a minimum waste of thread. One of the main problems was to obtain a continuous thread so that an almost unlimited number of stitches could be sewn rather than just being able to sew a small piece of material. Thimmonier's machine did not obtain continuous work, but Howe's (Fig. 7.6) and Singer's machines had a spool pin that the thread could be placed on so that the machine would pull on it at each stitch, the length of material to be sewn being proportional to the thread that was fitted on the spool.

The principles of sewing machines have remained unchanged to the present but some changes have been made to the automation of the processes in relation to the machine's efficiency and speed. Originally handle-operated, it is now pedal-operated so that the hands can remain free (in Fig. 7.6, the machine is operated by pulley V). Finally, sewing machines were designed to be driven by a foot-operated electric motor.

Howe's machine is operated by a cylinder C whose grooves serve as a guide for the main thread and the head for the hook. Figure 7.7 shows the head of the more

Fig. 7.4 The Jacquard loom

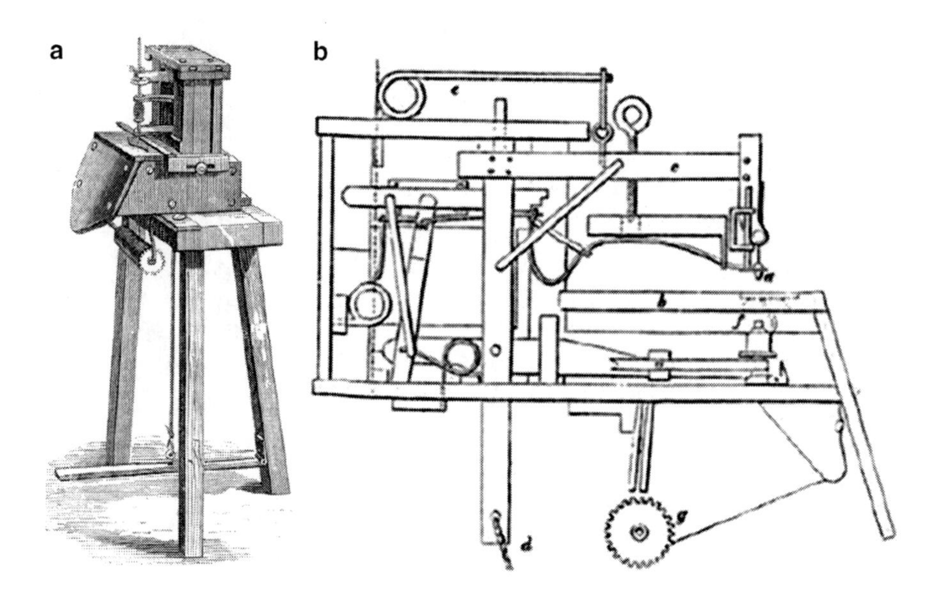

Fig. 7.5 The Thimonnier sewing machine [90]

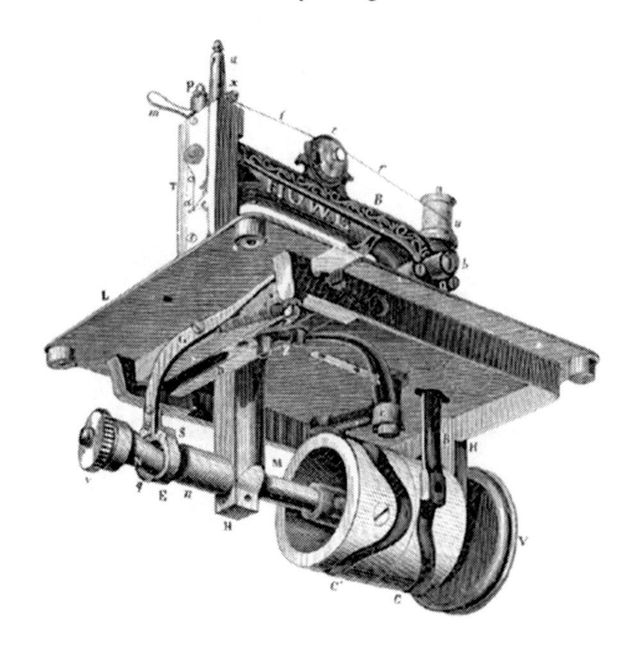

Fig. 7.6 Howe sewing machine [90]

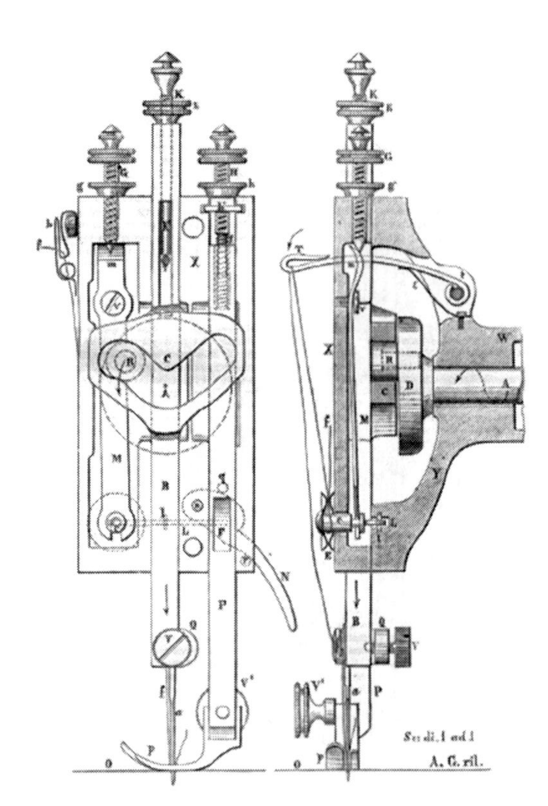

Fig. 7.7 The Singer machine head [90]

Fig. 7.8 Hemp and flax fibre softening machine [104]

advanced Singer machine. The thread is tensed by a hook, which can be positioned to obtain different types of stitches.

Frequently used materials were hemp and flax. In 1874, Salvá and Sanponts, both medical doctors, wrote a dissertation on an explanation and use of a new "machine for processing hemp and flax fibre" where there is a detailed explanation of the construction and features of their machine, by stressing all the benefits as compared to previous machines.

The machine, which is shown in Fig. 7.8, is driven by a horse which rotates a 72-tooth horizontal wheel. Then, this wheel drives a 12-spindle lantern which turns the grooved rollers whose job is to crush and stretch the flax and hemp. The operator's job consists in feeding the hemp and flax fibres through the rollers; then, he collects them, as can be seen in the illustration. The grooved rollers were located so that they remained separated when no flax was fed through them in order to avoid grinding or wear. In addition, when they were working, they could come together to press the flax or hemp as tightly as possible for optimum results.

Referring to the benefits of their machine, the authors stated: "Our machine not only corrects the evident defects of the Fibre Softener, Cutter, and Mill, but also, at one and the same time, performs the three operations of softening the fibres, and cutting and milling the flax".

On the Evolution of Handcraft Manufacturing

In the previous chapters we have discussed how the saw machine, of great importance for mankind, evolved. The last examples recall the saw mechanisms designs by Francesco Di Giorgio or Jacobus Strada. During the Industrial Revolution, several innovations were proposed for this machine.

Figure 7.9 is a design by Berthelot, who in 1782 drew this illustration of a wood saw. It had an up and down movement that is produced with the aid of a dual rack pinion. This provided the saw with vertical movement while the wood was moved horizontally by a vertical wheel-pulley system with a rope that moved the platform where the trunk was installed.

The second is the saw by Forest de Belidor. He was a French engineer who worked in several fields, including war machines and mathematics.

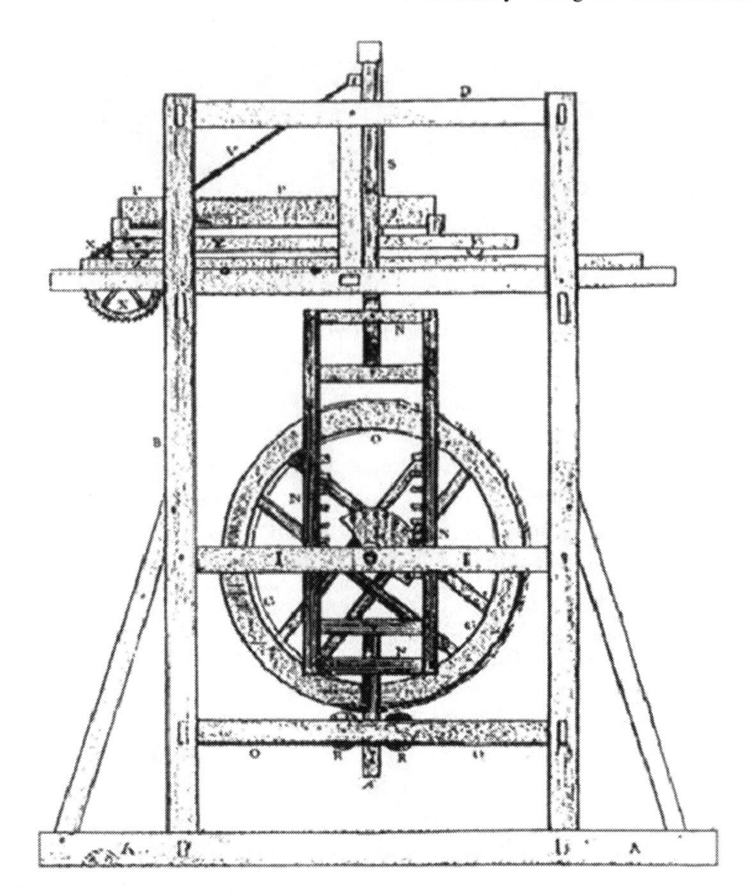

Fig. 7.9 Berthelot's wood saw [112]

Figure 7.10 shows a more advanced saw than Berthelot's, as it is dated 1787. It was also hydraulically powered but with the vertical movement that was transmitted by a crank and the horizontal forward movement of the wood was guided by three articulated rods and a wheel. This wheel had a smaller gearwheel that engaged the base supporting the wood so that when engaged, it provided the forward movement (right-hand side of the figure).

Only 5 years had passed between Berthelot's and Belidor's saws but the differences in appearance and design are considerable. The meticulous details in the drawing by Forest de Belidor indicate it is the result of an intense in-depth study of the mechanisms that was required to gain the best performance and productivity. This Belidor drawing makes Berthelot's saw seem rather crude and unpolished when the two drawings are compared. Just 1 year later, in 1788, Casado de Torres built a new saw but this time he made use of the newly invented steam engines to drive it.

In order to diversify the sawing work, Casado de Torres produced two saws, as shown in Fig. 7.11. In the first (on the left) he installed a series of parallel blades

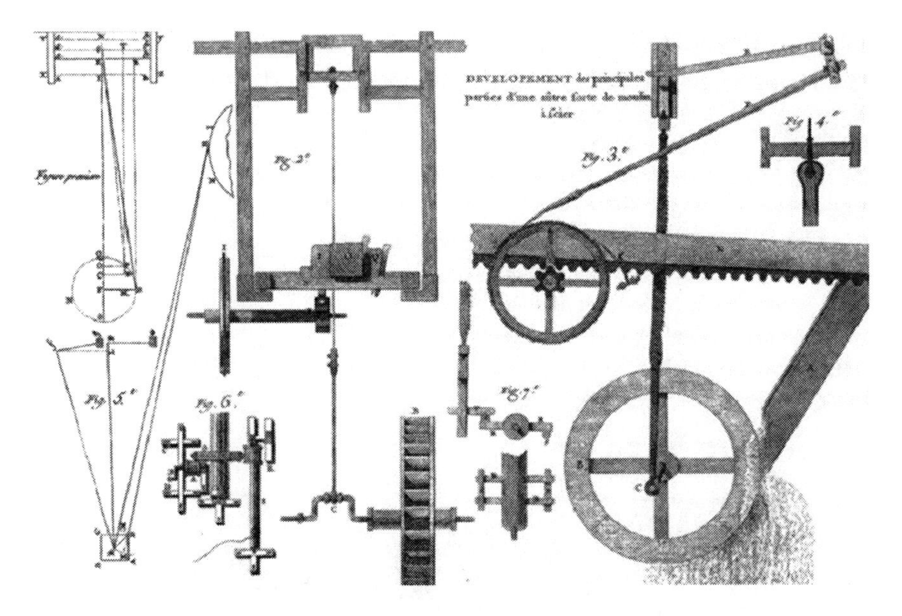

Fig. 7.10 Forest de Belidor's wood saw [112]

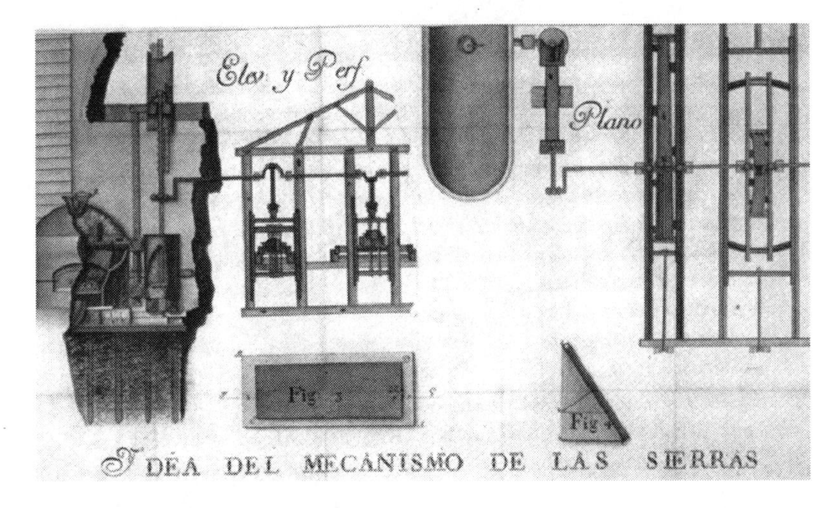

Fig. 7.11 Casado de Torres' wood saw [112]

to obtain strips of a thickness that could be obtained by the separation between the blades. In the second (on the left) he installed a turntable so that the wood could be turned to obtain curved saw cuts.

This is an excellent example of how to multiply the work by using only a single source of energy. The crank is connected to two rods, each belonging to a saw working in the opposite direction in order to balance the movement. The perfection and details of the drawing should also be noted, with diagrams showing a general view, elevation, and cross-section of the machine with as much information as possible, including exact geometric relations and dimensions so that the machine could be physically built without any assistance from the designer. Progress was experienced not only in the way how the source energy operates a machine but also in the way how the work is multiplied and increased in diversification of the end-products. Instead of one strip of wood, ten or more pieces could be cut and, if required, the wood could be turned to get a curved cut. Moreover, just one person was needed to supervise the operation and position the wood without any more effort required of them.

Figure 7.12 illustrates a press that was designed by J. Bramah in 1795. It had a dual mechanism where pressing was performed in a spring-loaded piston. This spring was, in turn, operated by a crank handle.

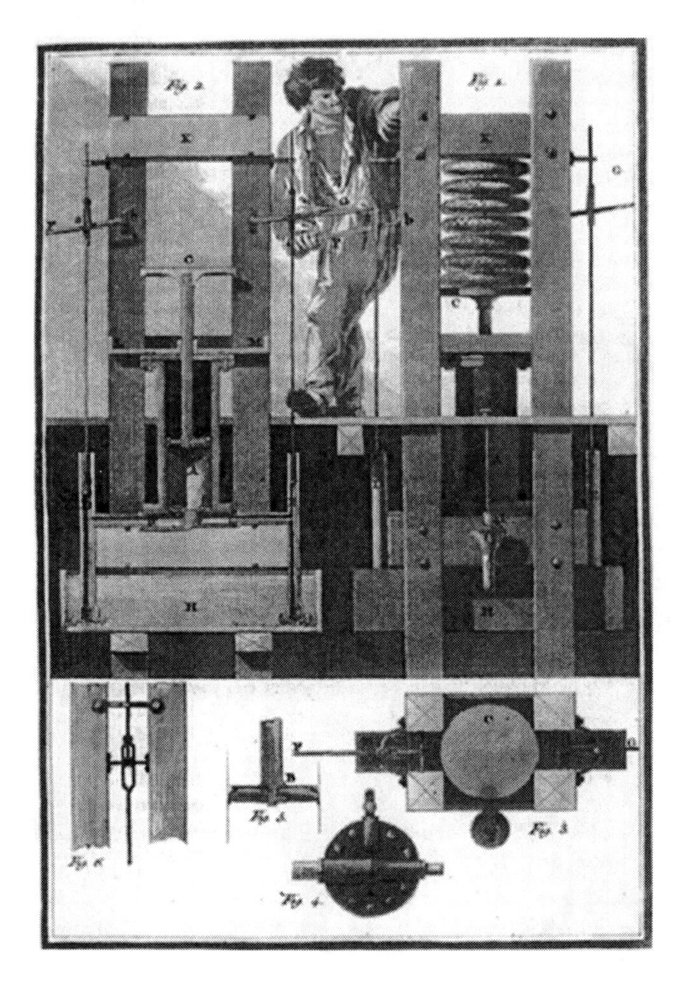

Fig. 7.12 J. Bramah's press [112]

On Machine Tools

The different types of saw and press machines previously illustrated are linked to the machine tools that were developed in the form of huge machines for the basic operations of cutting, drilling, grinding, and polishing.

The Industrial Revolution identified the need for a new kind of machine namely for mass production. Industrial machine tools had specifications and features that were needed for mass production and with long duration. The first machine of this type may be attributed to John Wilkinson who, in 1775, built a vertical drilling machine. Then, in 1794, Henry Maudslay developed an industrial mechanical lathe.

Figure 7.13 is a vertical lathe with three rotary heads that are useful to facilitate movement. The illustration shows the size of the machine compared to an operator.

Around 1860, the pages of machine tool books, magazines, and catalogues included lathes, milling machines, drilling machines, grinders, and boring machines. This gives some idea of how these types of machines had greatly evolved over a century.

Figure 7.14 illustrates a drilling machine, which is capable of making holes at an angle by partially rotating the tool. The hole height could also be changed. Even the entire body of the machine could be rotated as a characteristic that was extremely useful for large machines that were difficult to move. It is evident how machine tools increased in complexity over the years with the mechanisms that added accuracy, comfort, and efficiency to the machine operation and application goal.

Fig. 7.13 Vertical lathe [101]

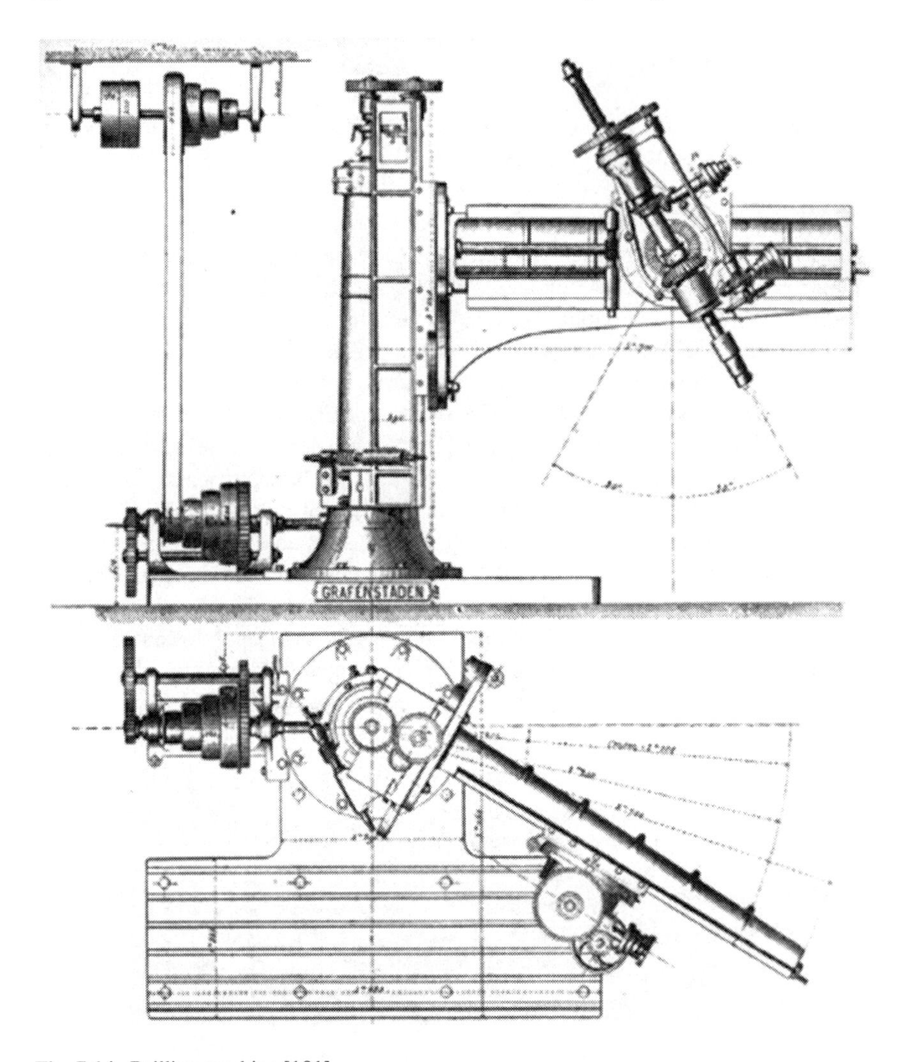

Fig. 7.14 Drilling machine [101]

On Hydraulic Machines

In previous chapters, successful hydraulic machines have been discussed as based on mechanisms to raise water to the required level. But further developments were proposed during the Industrial Revolution too.

One of the greatest design was by Louis XIV's engineers who produced the "Marly Machine", a huge civil engineering work, which in 1684 succeeded in supplying water to the Palace at Versailles by using 14 paddle wheels. Figure 7.15 shows a painting of the machine and Fig. 7.16 shows a cross-section of the paddlewheels. These wheels drove 64 pumps that raised the water almost 50 m. There was a second

Fig. 7.15 Part of a painting of the "Marly Machine" at Versailles

Fig. 7.16 Cross-section drawing of the "Marly Machine"

reservoir that contained another 69 pumps which again raised the water more than 55 m, and then a third stage of 78 pumps was used to raise the water up to the aqueduct that channelled the water to Versailles.

Figure 7.17 shows a bird's-eye-view of the set of wheels and pumps, which gives an idea of the overall size. Figure 7.18 gives a more detailed illustration of the action of a paddlewheel on the suction piston.

During the Industrial Revolution, the paddlewheels were replaced by much more efficient steam engines that took up much less room.

It is clear that during the Industrial Revolution this kind of machine benefited from the widening scope of mechanical know-how with the invention of machines that were more intelligent and with better technical specifications, like those in Fig. 7.19 showing Prunier's hydraulic system. It was built to supply the generators at the 1784 Vienna Universal Exhibition with water raising 14 m above the ground. In Fig. 7.19, the system is composed of two vertical suction pumps next to two

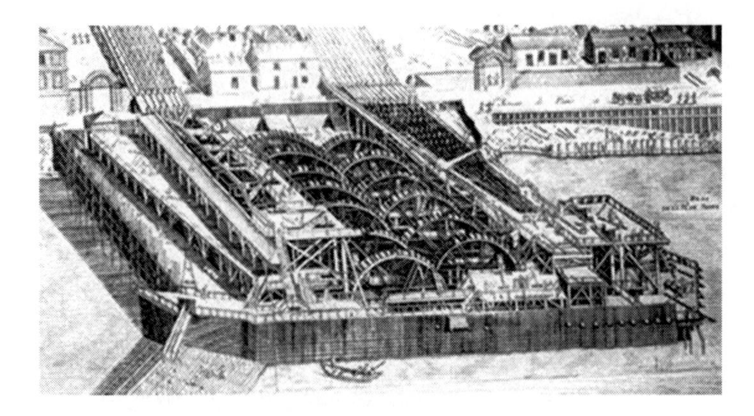

Fig. 7.17 Bird's-eye-view of the "Marly Machine"

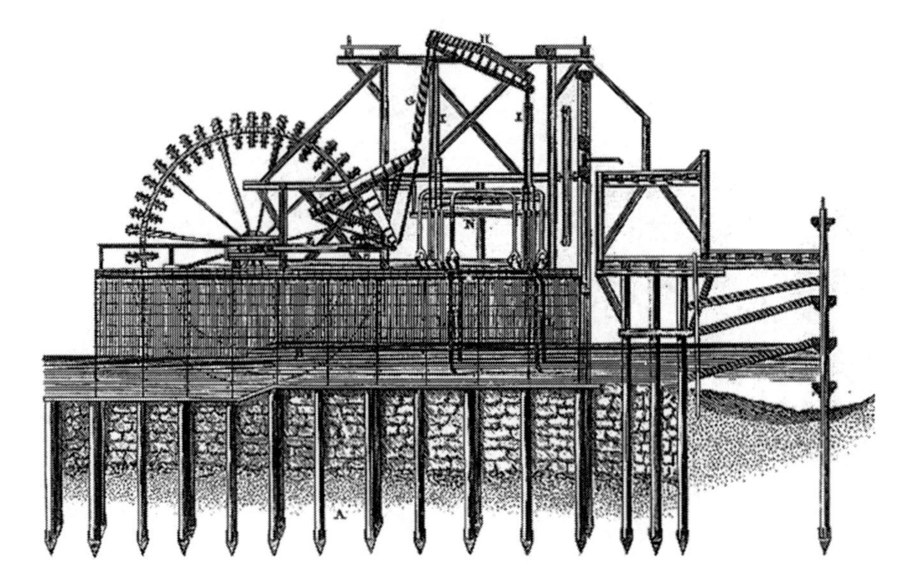

Fig. 7.18 Detail of the "Marly Machine"

steam engines. What made this machine original was its capacity to work by using both pumps together or, alternatively, to use a single but efficient mechanism.

In order to collect water from the Danube with this pumping system, 17-m long pipes were installed with 1 m diameter. In order to raise the water into the pipes, two pistons were used within which the two cones in the illustration were included. These cones had two rings, which let the water to flow down stream, and they closed the flow in the other direction of the piston up-stroke to retain the water. In the illustration, the elevator on the left is at the down-stroke position while the one on the right is completing the upstroke. The pumps were operated with a perfectly balanced alternate motion so that there could be a continuous flow of water from the centre pipe.

Fig. 7.19 Hydraulic machine at the Vienna Universal Exhibition, 1874 [6]

Let us now take a look at the history of turbines. This began with water-wheels in Egypt and China, but strictly speaking, the theory of water turbines began with Euler (1754) when the construction of Fourneyron (1827) was successfully achieved. Two type of installation for water turbines were considered, namely a horizontal wheel, as shown in Fig. 7.20a), and a vertical wheel, as shown in Fig. 7.20b).

On Steam Engines

The usefulness of steam is based on its performance. According to texts of the time, "A quintal (~46 kg) of coal carefully used produces an amount of work that is greater and more regular than that of 12 strong, tough men". This statement makes

a

b

Fig. 7.20 (**a**) Horizontal wheel turbine from the Vienna Universal Exhibition, 1874 [6]. (**b**) Vertical wheel turbine, 1890 [105]

it quite clear that the steam engine and the use of coal as an organic propulsion material had become an absolutely necessary aid.

As already stated, the first sketch of a steam-driven machine was by Hero of Alexandria and it was not until 1606 that this method again began to be studied by Jerónimo de Ayanz, and by Giovanni Branca in 1629.

In 1673, Denis Papin (1647–1712) and Christian Huygens (1629–1695) proposed a steam engine, but it was Thomas Savery (1600–1715) in 1689 who used steam to extract water from the mines with his "miner's friend" that is shown in Fig. 7.21.

However, Savery's engine had several drawbacks: the engine was located about 12 m underground and the pressure reached ten atmospheres, which meant the pipes had frequent failures. Nevertheless, the illustration shows Savery's principal, which was that vacuum was created in the containers for raising the water without the aid of pistons.

When the complications of Savery's engine were clearly experienced, Newcomen studied them and designed a new engine, as shown in Fig. 7.22, which he revealed in 1712. This used a piston that was moved by the vacuum that was created by the

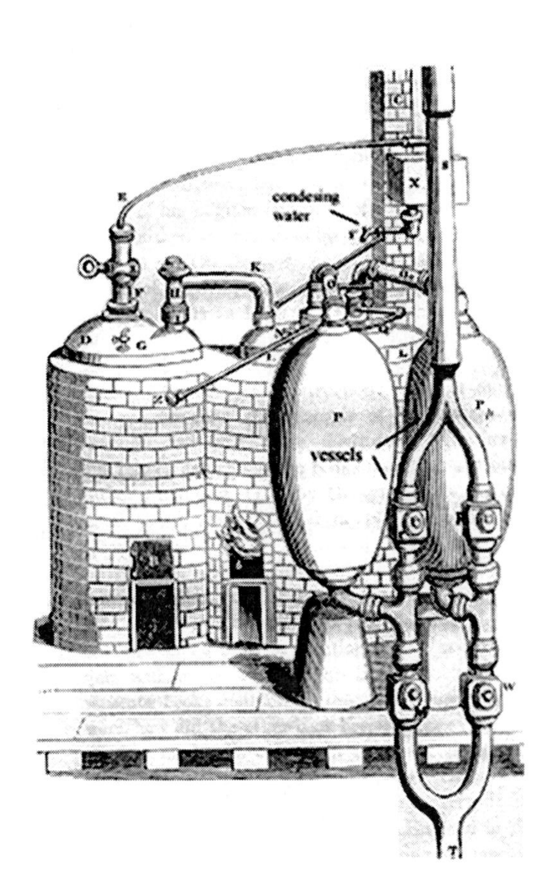

Fig. 7.21 Savery's steam engine

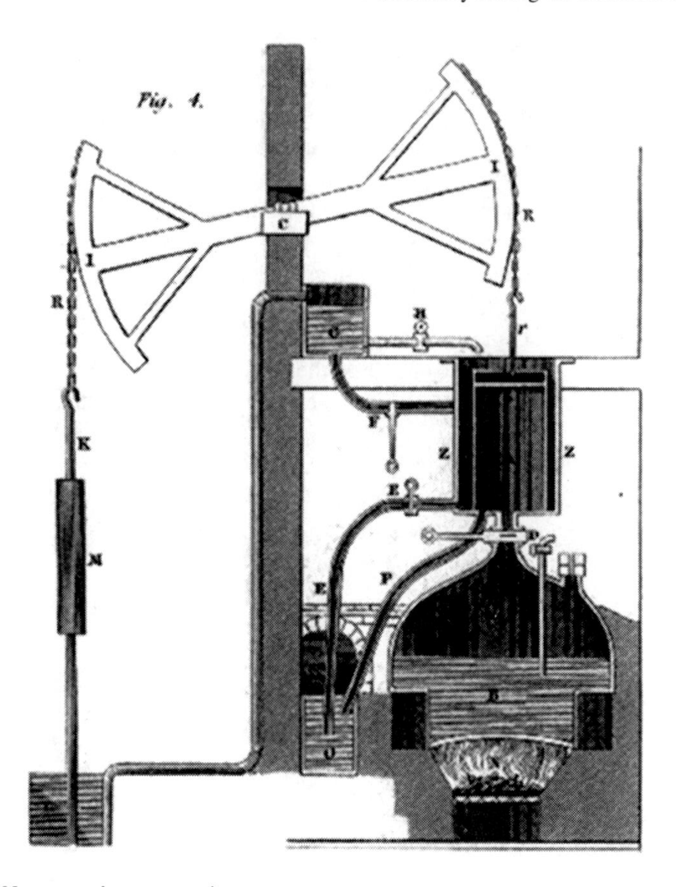

Fig. 7.22 Newcomen's steam engine

condensation of the steam. This piston moved the large pulley to the left that raised the water. This machine could reach 12 strokes per minute with the aim of raising 45 L of water from a depth of 46 m at each stroke.

The ultimate steam engine was designed by James Watt (1739–1819). In 1769, he conceived an engine that was a considerable improvement on Newcomen's design, since it reduced the steam losses and increased mechanical performance. The way to achieve this was simple: Watt condensed water in another cylinder that was distinct from the drive piston, by connecting both with a pipe and by covering the piston with a steam jacket to preserve the heat. Watt added other improvements with the centrifugal regulator by using balls for speed control and by obtaining rotary motion instead of linear motion, as shown in Fig. 7.23, where the engine turns the wheel on the left by means of the articulated rocker arm at the top with a connecting rod that moves a gear assembly at its lower end to connect with the drive wheel. This planetary gear was one of the five solutions that were patented by Watt to obtain circular motion as machine output.

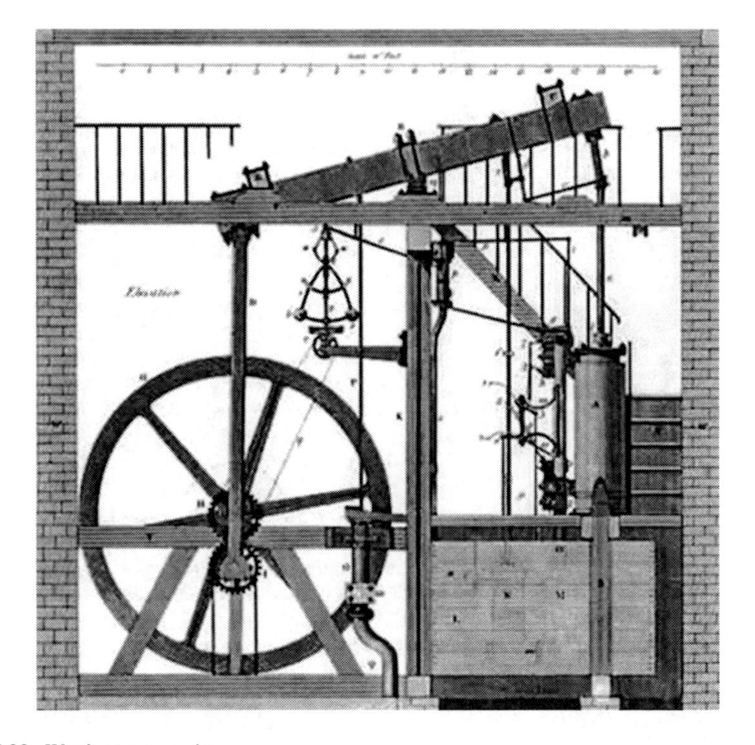

Fig. 7.23 Watt's steam engine

Watt continued working on his engine by making more improvements. He owed part of his success to Wilkinson, who in 1775 invented a cast-iron drilling machine that was precise up to 1 mm. This meant that heat losses from the steam engine were greatly reduced and Watt began to propose building a double-acting engine. With this success, Watt began to pride himself on his new invention: the so-called "Watt's singular mechanism" and its later extension "Watt's extended mechanism". This mechanism consisted of an articulated four-bar linkage that can be seen in Fig. 7.24, as a drawing of one of Watt's double-acting engines by Agustín de Betancourt. The articulated mechanism makes the piston to generate the rotary motion of the wheel on the right during both the upstroke and downstroke piston action.

British progress also reached Spain, but somewhat later. In 1813, Jorge Juan designed the single action steam engine as shown in Fig. 7.25. The illustration shows it to be a water-raising device as based on straight-line vertical motion.

From 1850, the main concern was to increase engine power, which gave rise to various constructions, like O. Patin's whose advertisement for his engine is shown in Fig. 7.26, as it appeared in machine magazines in 1860. Machines like the one in Fig. 7.27 were also built with triple-expansion or smaller engines for work requiring smaller power.

When the steam engine had reached the above successful general developments, engineers were attracted to individual parts design or constructional details in order

Fig. 7.24 Watt's double acting steam engine [112]

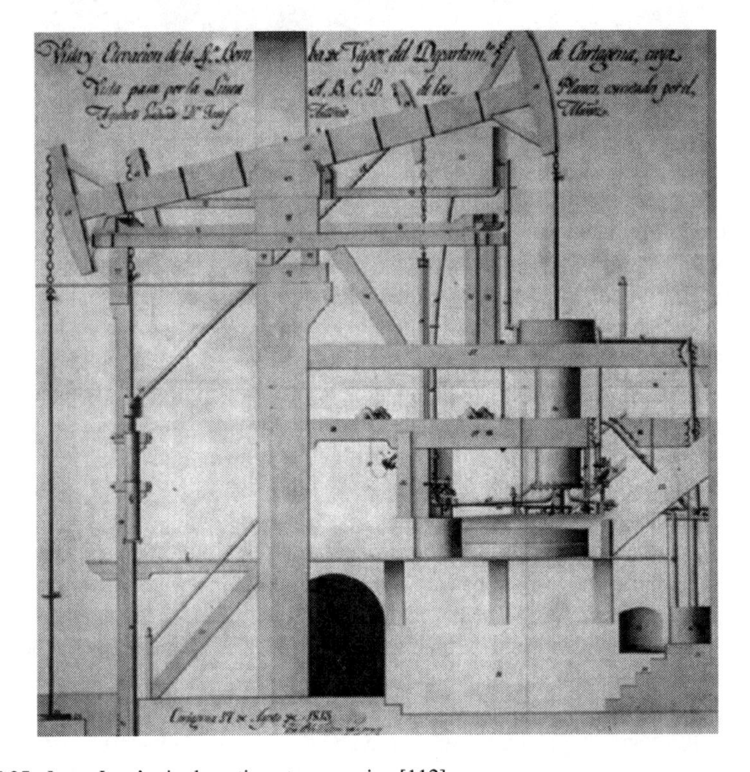

Fig. 7.25 Jorge Juan's single-acting steam engine [112]

Fig. 7.26 O. Patin's steam engine advertisement [101]

Fig. 7.27 Triple-expansion steam engine [106]

to add some improvement or innovation. This led to many layouts and changes that were experienced as shown in the example of Fig. 7.28 in which a steam engine is proposed with two pistons instead of one with a V-shaped layout to save space. This engine was built in 1879, over 100 years after Watt had built his double-acting engine. The illustration shows an amazing development in the technology of the engine as well as in aesthetics and size. It should not be forgotten that the

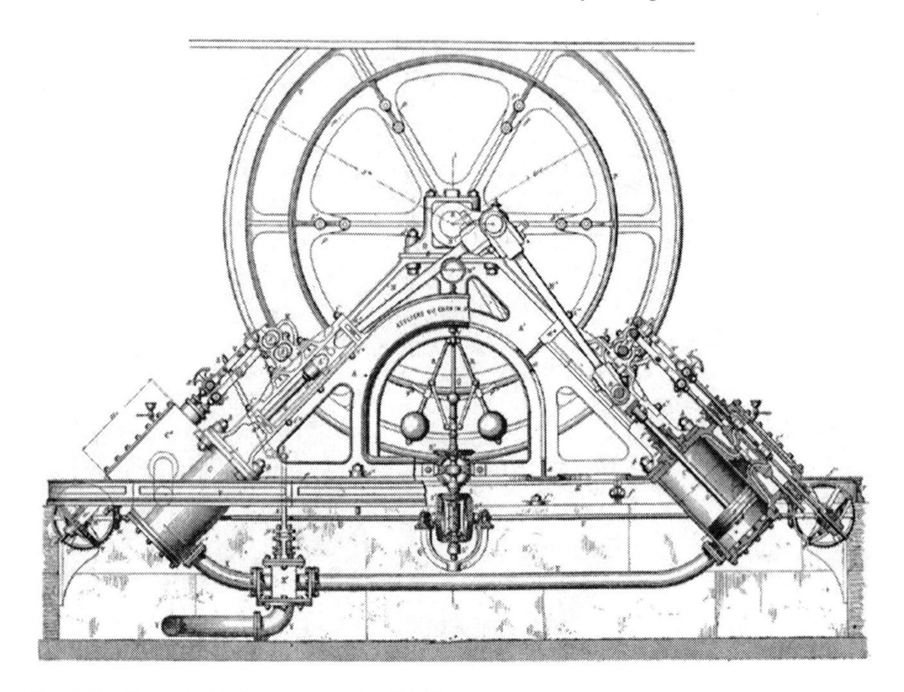

Fig. 7.28 Slanted cylinder steam engine [106]

improvement in the different engine parts was aided by the improvement in machine tools, as previously mentioned.

Those developments gave a great variety of engines that tried to excel in the different qualities demanded by consumers. After the steam engine came onto the scene, one of the priorities was to reduce its size to make it "domestic". There were many families that could use a steam engine at home to do the harder jobs, so that the children or women could supervise the work while the men worked outside the house.

Thi is the case of the vertical steam engine in Fig. 7.29. Its inventors (The Buffaud Company in Lyon, 1874) made it in this configuration so that it could be moved around the house, since the boiler could be rotated. Moreover, the boiler was small and the water pipes were installed so that they could be easily replaced.

Among the engines that were exhibited at the 1874 Vienna Universal Exhibition, the one in Fig. 7.30, that was manufactured by Frederick Siemens, attracted the greatest interest. This engine design introduced a radical change as compared to Watt's conception of a steam engine, since it had no piston connecting rod or crank and its external appearance was more like that of a screw. The innovation consisted in the engine's rotary motion and at no time did the steam escape to the outside. The engine was started by filling the bottle labelled A with water through an hole in the neck which was closed once the water was inside. Then, the lower part of the bottle was heated and the steam rose up through the spiral, forcing the bottle to spin.

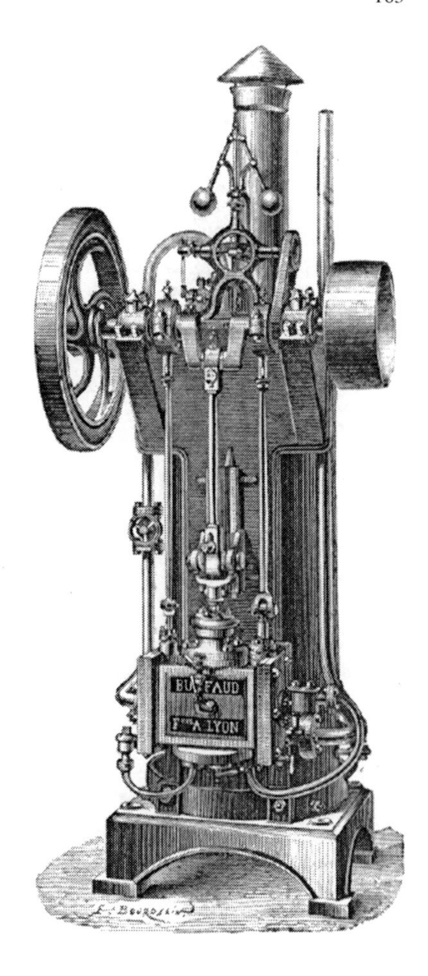

Fig. 7.29 Vertical steam engine [6]

The steam then passed to the coil (labelled *C*, the spirals of which were designed in the opposite direction to those of the bottle-spiral) and expelled the air through *k*. (Immediately after the air had left the spiral through *k*, it closed.) When the system was closed, the condensed steam in the spiral fell to the bottom of the bottle, while the gases produced by heating the water passed through pipe *d* due to the action of the refractory flange *B*. This simple way of obtaining rotary motion as supplied by the bottle shaft *L-T*, was defended by its inventor not only for its simplicity but also for its minimum fuel consumption by making use of all the steam expansion force or its direct action, which avoided leakage or cooling.

The performance of steam engines was improved and power consumption was reduced and ever more factories and builders sprang up, but instead of specialising in building complete engines, they did research and introduced innovations in the parts.

One of the mechanisms that was invented by Watt was the speed regulator using balls which kept the speed more or less constant by means of centrifugal force.

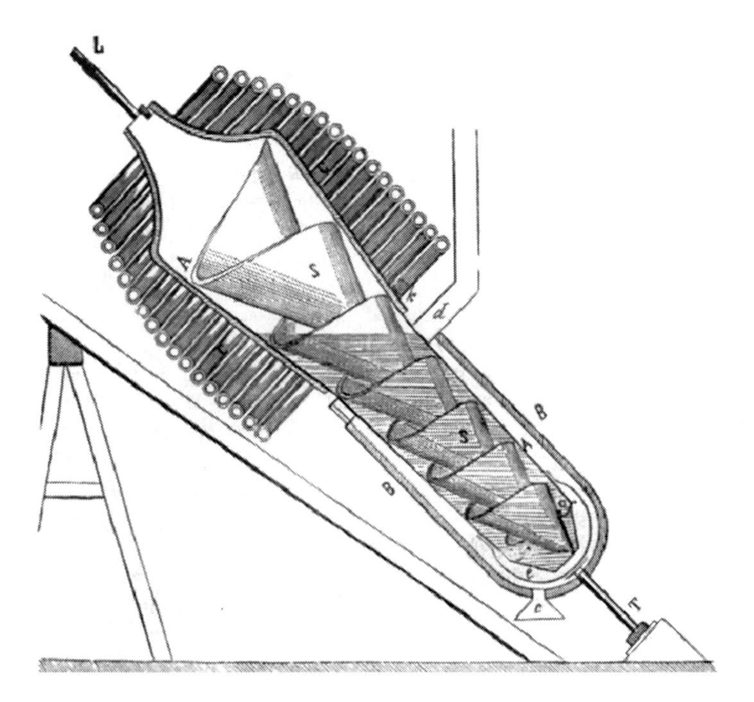

Fig. 7.30 Siemens' steam engine [6]

Figure 7.31 shows a new solution and innovative ball regulator. It also uses balls but these are aided by springs which are connected to the springs, to act as a centrifugal force for a larger or smaller pull on the spring which increases or decreases the speed of the wheel to which the regulator is installed.

On the Development of Transport

The beginnings of the railway can be dated to 1802, when Richard Trevithick patented a high-pressure steam engine to drive a locomotive. Since then steam locomotives evolved very rapidly. In addition a new means of transport arose in towns and cities: the car. Those machines were developed as much as based on mechanism designs as shown in the examples in Figs. 7.32 and 7.33.

On Automatic Astronomical Devices

In this field, we cannot fail to mention "Herschel's Great Telescope". There is a letter by Herschel from 1796, where he accepted a commission to build two telescopes for the King of Spain. The telescope was completed a year later and it was

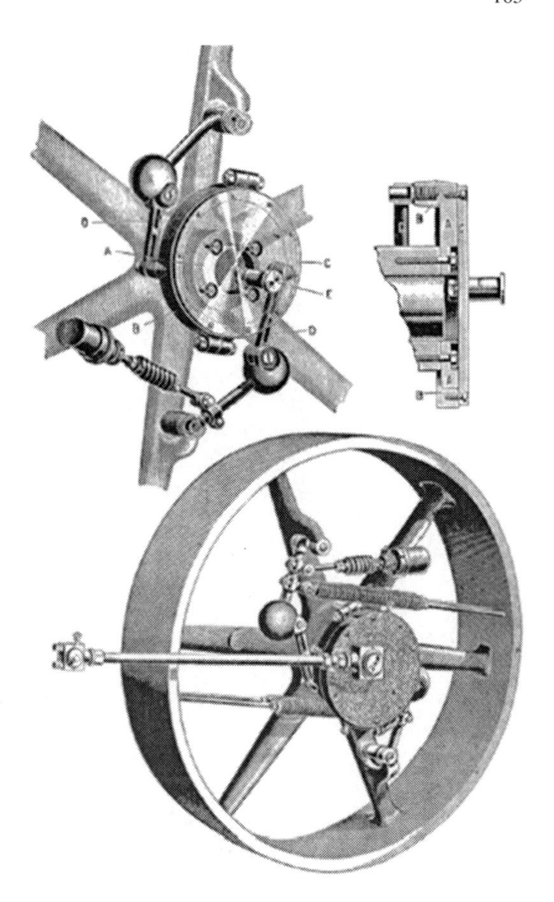

Fig. 7.31 May regulator [106]

installed in the Royal Observatory in Madrid. It was two feet in diameter with a 25 ft focal length, and it was considered the best of its time. Thus, the observatory became part of an up-and-coming branch of astronomy called cosmology and was equipped with one of the world's best telescopes: number one in optical quality in Herschel's opinion and second in size.

Figure 7.34, from the paper entitled "Industrial Archaeology: From the seventeenth to the twenty-first century. Reconstruction of Herschel's Telescope", is indicative of the size of the mechanical device of which the telescope was part. It was supported on a unique circular base-frame comprising oak columns and beams. The tube was suspended at one of its ends by a block and tackle and supported at its highest point on a crossbeam, the other end being supported on the base. The observer could rotate the whole frame as well as vary the inclination of the tube, which were the main movements possessed by the telescope. Figure 7.35 shows some of its mechanisms.

We also have evidence of the improvements the inventor made to the mirrors, enabling him to observe the planet Uranus that had been discovered only a few

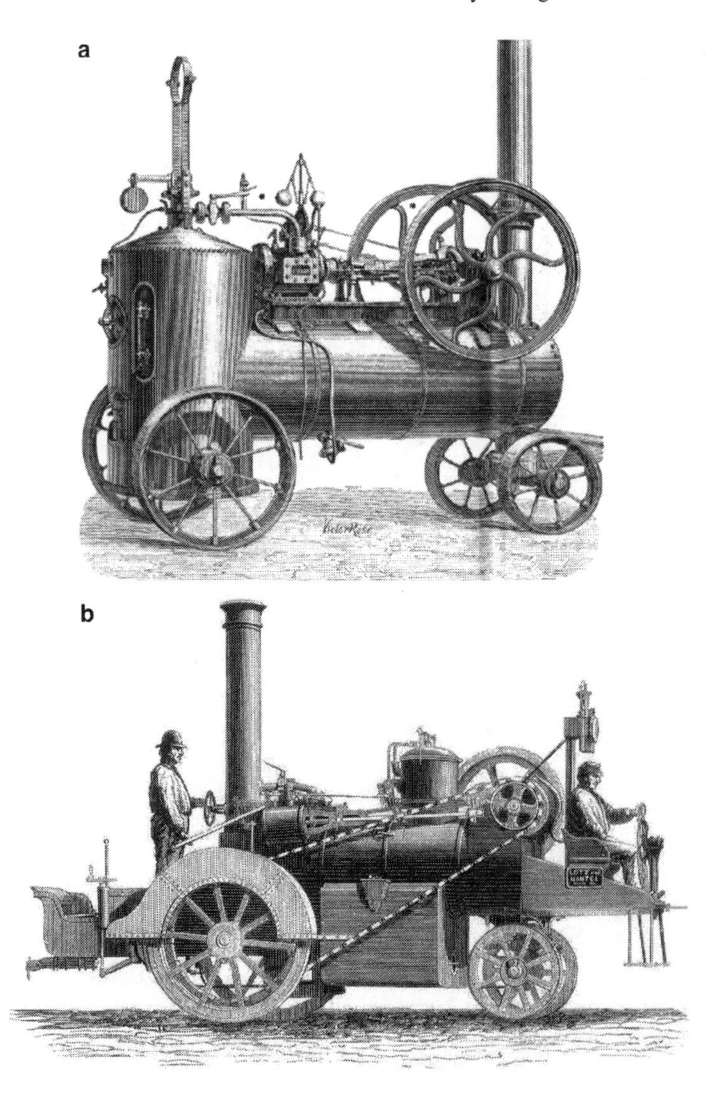

Fig. 7.32 (**a**) 1874 locomotive [6]. (**b**) 1864 Steam vehicle [105]

years earlier. The mechanism incorporated an automatic star tracking system operated by a large clock.

Unfortunately, this telescope was destroyed during the War of Independence in 1808 with just a few scarce remains surviving and a picture of its structure, thanks to some magnificent plates prepared by a naval officer called Mendoza.

Fig. 7.33 Motor velocipede, 1890 [105]

Fig. 7.34 Herschel's telescope [18]

Fig. 7.35 Detail of some of Herschel's telescope's mechanisms [18]

Chapter 8
A Vision on Machines

In the western world, the practice of machine construction and the theory of mechanics came together at the end of the Renaissance with two lines of development: treatises in the form of rationally classified machine collections and machine studies as an application of mechanical physics.

Practical activity was developed through collections of machine design in parallel with an interest in machine-applicable theoretical aspects that initially sought to recover the knowledge of Antiquity.

Thus, Greek technology and the machines of Roman engineers were again discovered and examined, at first out erudite and humanistic interest.

These early studies on machines attracted the interest of many researchers in the fields of philosophy and mathematics. However, this interest was still far from applying the knowledge to a professional use of the rediscovered mechanical devices.

The most outstanding personalities in this development of a theoretical discipline on machine design may be considered to be firstly Francesco Di Giorgio and Agostino Ramelli, and then Guidobaldo Del Monte and Galileo Galilei.

Francesco Di Giorgio wrote a treatise to explain the differences between a series of machines and how they can work by referring mainly to his own experience. This was a first specific handbook on machines (Fig. 8.1). These kinds of ordered notes were made by several authors, as shown in the example in Fig. 8.2 which was inspired by the work in Fig. 8.1.

On Re-examining Greco-Roman Works

The recovery of Ancient Classical culture on mechanics occurred through the discovery of the works of Greco-Roman authors which were translated, interpreted, and illustrated in an attempt not only to understand the mechanical problems of the past but also with the purpose of finding suitably rigorous and rational solutions.

The illustration by Cesare Cesariano of a machine described by Vitruvius (1521) in Fig. 8.3 is an example. It is a graphic interpretation that follows the rules of

E.B. Paz et al., *A Brief Illustrated History of Machines and Mechanisms*,
History of Mechanism and Machine Science 10, DOI 10.1007/978-90-481-2512-8_8,

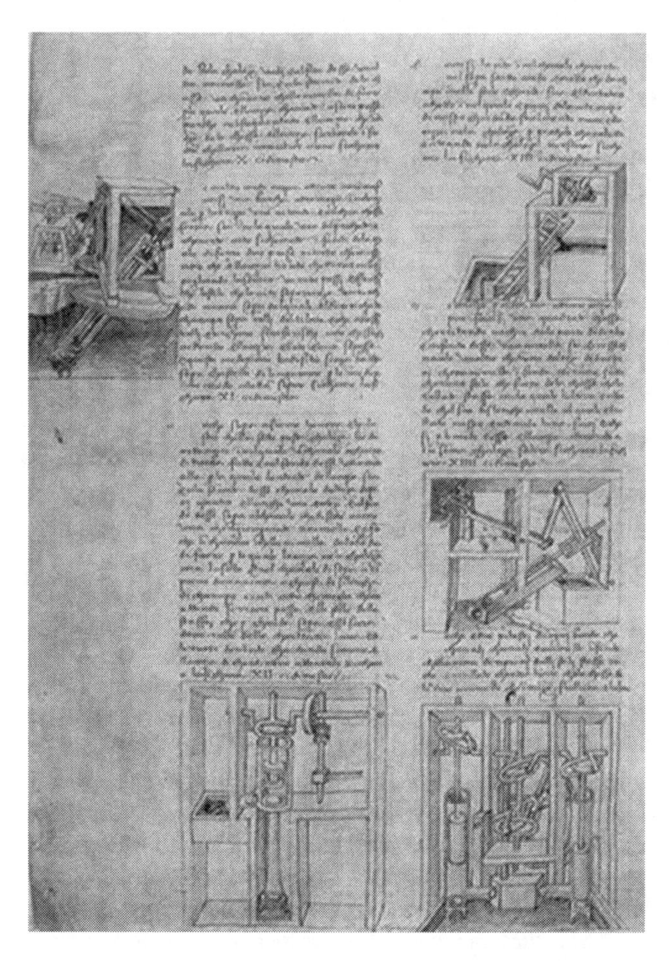

Fig. 8.1 Treatise on pumps by Francesco Di Giorgio (1439–1501)

drawing during the Renaissance by showing the several rope and pulley devices that were described by Vitruvius for application to cranes.

Chapter 5 of this book looked more deeply into how the Renaissance promoted the study of machines, starting from a reconsideration of antique machines.

The erudite form of machine collections were rapidly expressed as catalogues with updated comments on the use of the machines, such as the "Theatrum Machinarum" by Agostino Ramelli, first published in printed form in 1588, as shown in Fig. 8.4.

Book printing gave an enormous impulse to the dissemination of these treatises, whose authorship now contained signs of intellectual property, as a basis of present-day patents.

The tradition of machine catalogues has evolved over the centuries right up to modern-day handbooks and patents, preserving certain Renaissance features.

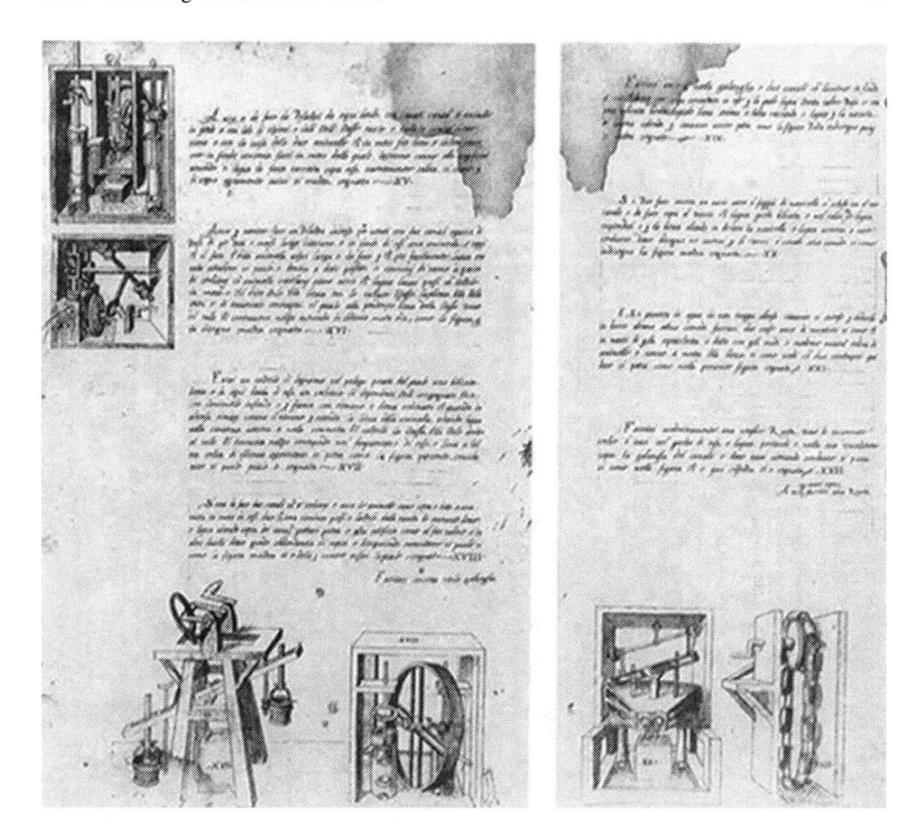

Fig. 8.2 Notes on pumps by Bartolomeo Neroni (Il Riccio)

The rediscovery of Greek mechanics stimulated design theory and machine function to be approached as a discipline that was for practical use. Initially, the ideas of Aristotle, Archimedes, and the Greek geometers were recovered and then interpretations and studies of the mechanics of the machines were developed in an original way and with modern analytical aspects.

This is the case of the milestone work by Guidobaldo Del Monte. Using Archimedes' approach, he wrote his erudite work on the mechanics of elementary machines in "Mechanicorum Liber", that was published in Latin in 1577 and then as "Le Mechaniche" in Italian in 1581 (Fig. 8.5).

Continuing along these lines, like many others who tried to propose ever more mathematical rigor, Galileo Galilei not only dealt with the subject of machine mechanics with greater clarity and a better overall vision for their practical applications, but he also discussed the subject as part of his academic teaching. This was probably the first course that independently approached the subject (see the illustrations in Fig. 8.6a from Galileo Galilei's French edition "Les Mécaniques", 1634).

By improving the concepts that were elaborated on by Del Monte, Galilei proposed the mechanics of the lever as a basic principle for modelling and explaining the operation of all elementary machines. He developed a very early modern

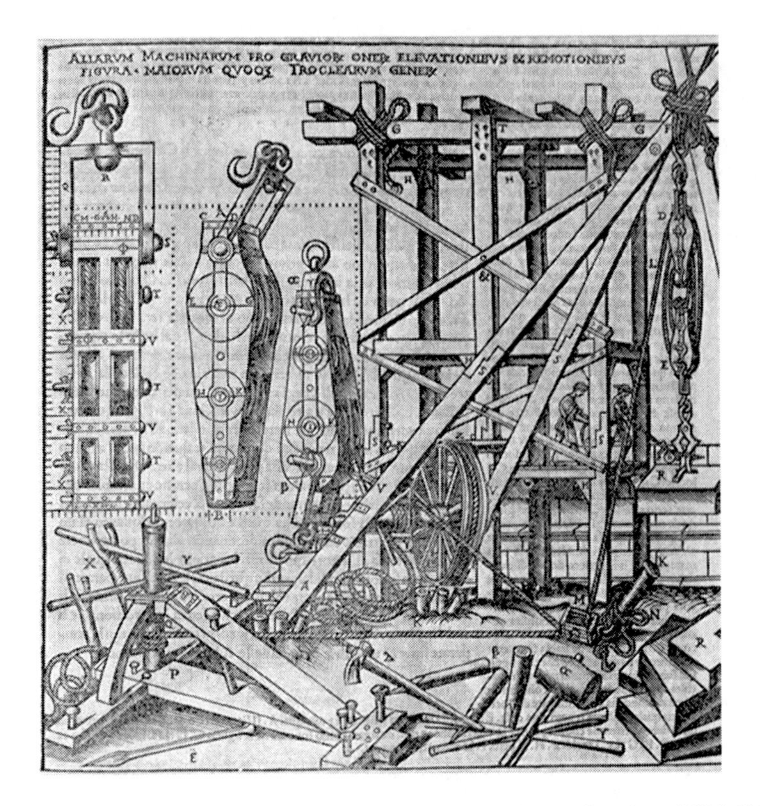

Fig. 8.3 Treatise by to Vitruvius on Roman devices according Cesare Cesariano (1521) [25]

approach with kinematic diagrams and basic concepts for studying dynamic equilibrium in mechanical systems.

With Galilei's work, machines were considered objects worthy of study inside academic spheres for teaching and research. Also remarkable are Leonardo da Vinci's classifications in Fig. 8.6b not only of machines in general but also of their component parts such as types of springs, joints, gears, etc.

On the Systematisation of Machine Study

After the machine treatises of the Italian and European Renaissance and "The Twenty-One Books of Devices and Machines" (see Chapters 5 and 6), the Industrial Revolution brought books on mechanisms that combined illustrations and texts showing the advances of that time.

The outcome of that period was the development of new machines with greater capacity and power. At the same time, an enormous stimulus was provided for a large range of studies on structure and functionality by machine scientists and research specialists in mechanical engineering.

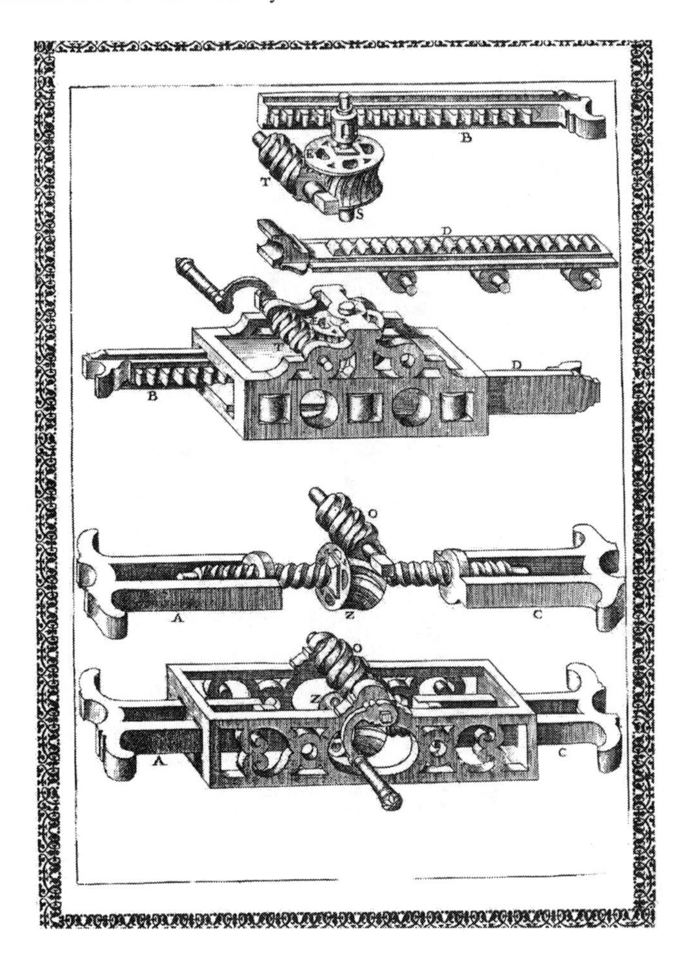

Fig. 8.4 Examples from the machine collection of Agostino Ramelli in 1588 [93]

As never before in the past, this interest in studying at a scientific level also led to the establishment of a considerable community of experts and inventors who contributed, often anonymously, to technological and social developments.

The need for professional figures in mechanical engineering was established from the beginning of the Industrial Revolution, regardless of any training or a military career (as had been the case in the past)

Therefore, in 1794 the École Polytechnique was founded in Paris as an academic structure for training engineers, who at that time devoted their time mainly to structures of civil engineering and industrial machinery.

In many other countries, and not only highly industrially developed ones, this need was identified and solved with suitable formation structures. For example, schools for engineers were founded in Sao Paulo (Brazil) and in Mexico City (Mexico) in the same decade. But since the beginning the École Polytechnique played an important

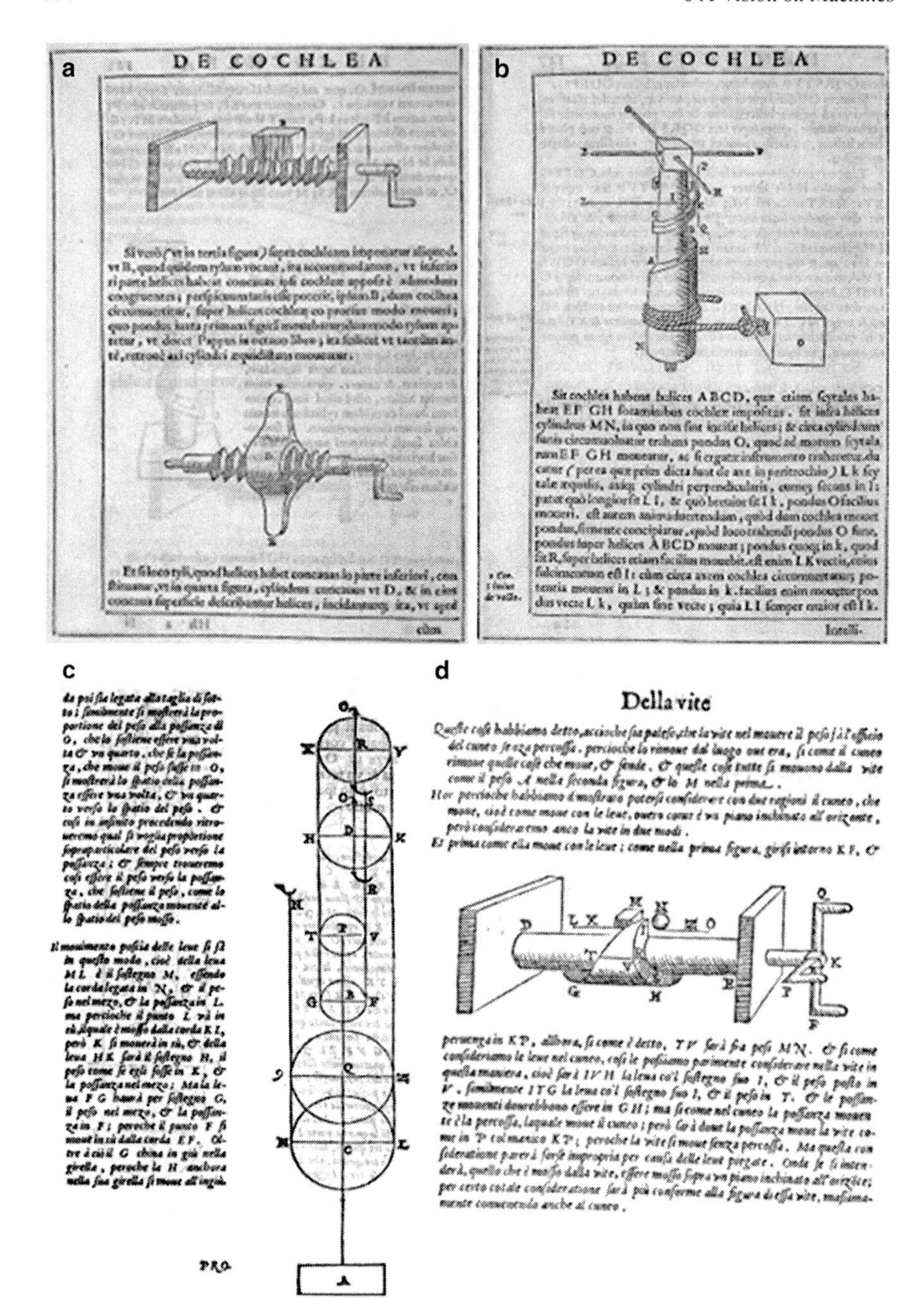

Fig. 8.5 Illustrations from the texts [82,83]: (**a**) "Mechanicorum Liber" in 1577; (**b**) "Le Mechaniche" in 1581

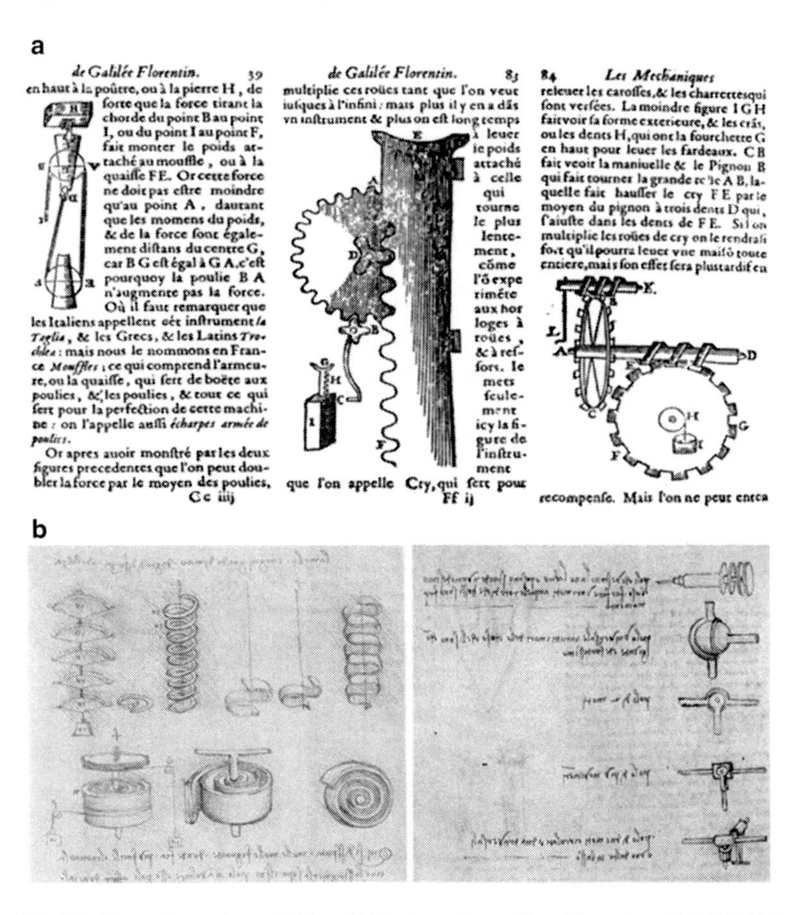

de Galilée Florentin. 39

en haut à la poûtre, ou à la pierre H, de sorte que la force tirant la chorde du point B au point I, ou du point I au point F, fait monter le poids attaché au mouffle, ou à la quaisse F E. Or cette force ne doit pas estre moindre qu'au point A, dautant que les momens du poids, & de la force sont également distans du centre G, car B G est égal à G A, c'est pourquoy la poulie B A n'augmente pas la force. Où il faut remarquer que les Italiens appellent cét instrument *la Taglia*, & les Grecs, & les Latins *Trochlea*: mais nous le nommons en France *Mouffles*; ce qui comprend l'armeure, ou la quaisse, qui sert de boëte aux poulies, & les poulies, & tout ce qui sert pour la perfection de cette machine: on l'appelle aussi *écharpes armée de poulies*.

Or apres auoir monstré par les deux figures precedentes que l'on peut doubler la force par le moyen des poulies,

Cc iiij

de Galilée Florentin. 83

multiplie ces roües tant que l'on veut iusques à l'infini: mais plus il y en a dās vn instrument & plus on est long temps à leuer le poids attaché à celle qui tourne le plus lentement, cōme l'ō experimēte aux horloges à roües, & à ressors. Ie mets seulement icy la figure de l'instrument que l'on appelle Cry, qui sert pour

Ff ij

84 *Les Mechaniques*

releuer les carosses, & les charrettes qui sont versées. La moindre figure I G H fait voir sa forme exterieure, & les crās, ou les dents H, qui ont la fourchette G en haut pour leuer les fardeaux. C B fait veoir la manivelle & le Pignon B qui fait tourner la grande roüe A B, laquelle fait hausser le cry F E par le moyen du pignon à trois dents D qui, s'aiuste dans les dents de F E. Si l'on multiplie les roües de cry on le rendra si fort qu'il pourra leuer vne maisō toute entiere, mais son effet sera plus tardif en recompense. Mais l'on ne peut entca

Fig. 8.6 (a) Illustrations from Galileo Galilei's work as "les Mecaniques" 1634 [44]. (b) Classification of types of springs and joints by Leonardo da Vinci [133]

role and was prominent as a centre for engineering development, since a large number of top scientists were involved in the École's activities.

A need for industrial engineers was identified for a rigorous academic training. Thus, at the beginning of the nineteenth century, several European universities had a "curriculum" for engineers where, among the compulsory courses machine technologies and the kinematics of mechanisms were given as fundamental.

Monge's approach that was based on the application of descriptive geometry to the study of mechanisms, inspired many other works during the whole nineteenth century. In particular, Agustín Betancourt and José María Lanz are examples, who, by using a kinematic analysis, approached the kinematics of mechanisms as the "study of mechanical systems without taking account of the causes that produce their motion". In their "Essay on Machine Composition" (1808), based on Monge's ideas and plans for teaching mechanisms, they made a detailed study of types of movements and classified them according to how the movement can be transformed by the variety of mechanisms. Thus, they created the first modern systematic machine treatise, which will be studied throughout the twentieth century too. Figure 8.7 shows an illustration from this text.

The classification by Betancourt and Lanz is structured as ten categories starting from four types of movements: uniform rectilinear, alternate rectilinear, uniform circular, and alternate circular. The relationships between these four point movements and their transformation into each other give ten categories. Figure 8.8 illustrates a part of the summarizing table for this classification.

The importance of a theory of mechanisms was recognized as so fundamental that in 1831 it became a subject in its own right and was identified with the specific name of kinematics of mechanisms with the aim of denoting the study of the motion of a mechanism regardless of the actions causing it. This was followed by the science of mechanisms named the theory of machines and mechanisms (TMM) which rapidly spread throughout Europe.

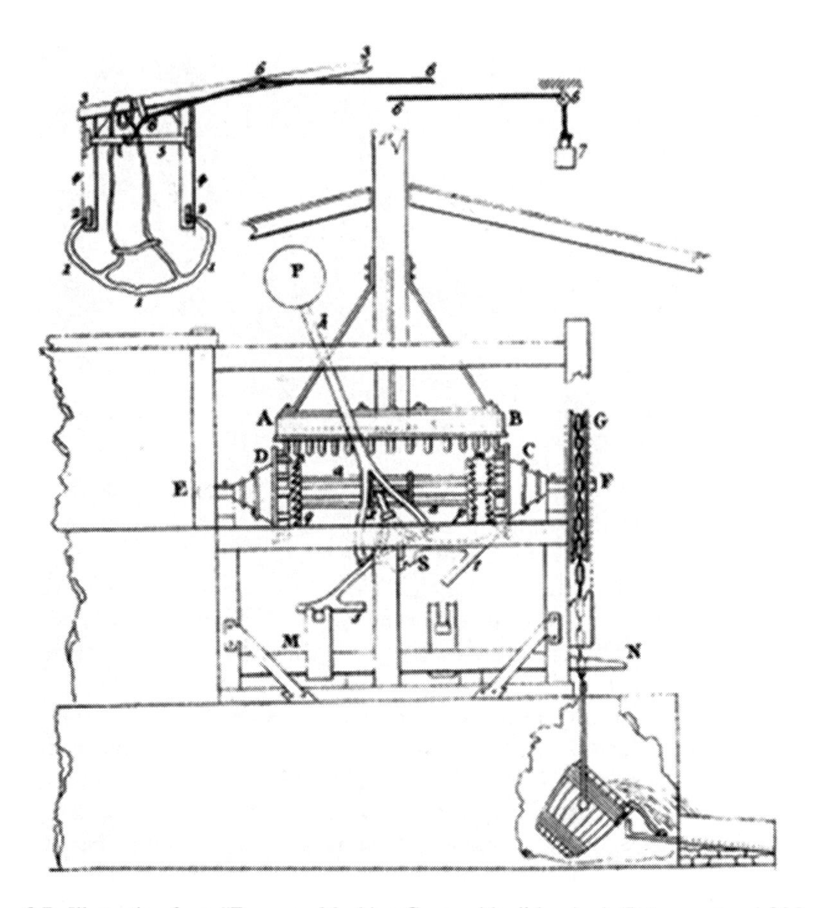

Fig. 8.7 Illustration from "Essay on Machine Composition" by A. de Betancourt and J.M. de Lanz [73]

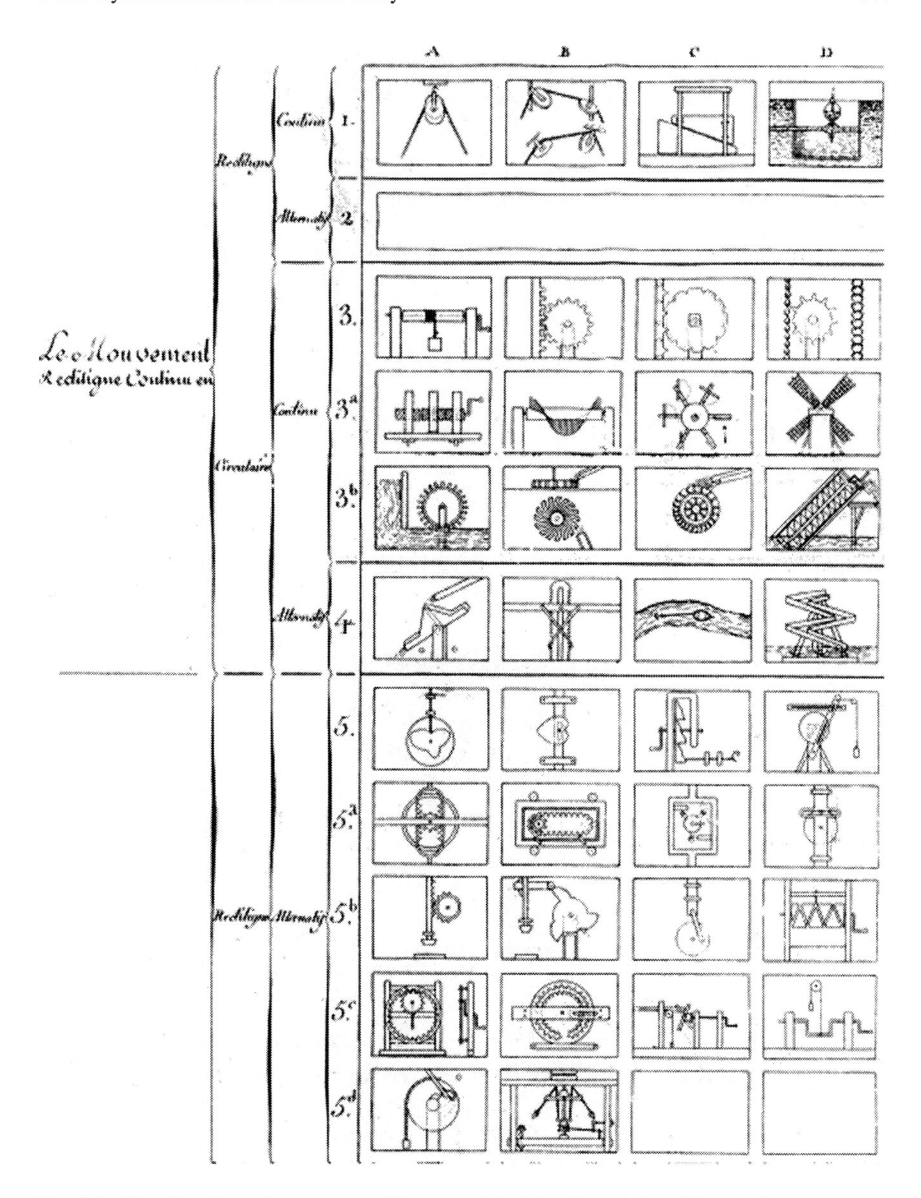

Fig. 8.8 Classification of movements "Essay on Machine Composition" by A. Betancourt and J.M. Lanz [73]

After the works by Betancourt and Lanz, M. Hachette followed the same model in his book "Traité élémentaire des machines" in 1811, since he classified mechanisms as function of movement capability into the same ten types. Figure 8.9 shows a reproduction of the title page of this text.

TRAITÉ ÉLÉMENTAIRE

DES MACHINES,

PAR M. HACHETTE,

INSTITUTEUR DE L'ÉCOLE IMPÉRIALE POLYTECHNIQUE.

PARIS,

J. KLOSTERMANN fils, Libraire de l'École Impériale Polytechnique, rue du Jardinet, n°. 13, quartier St.-André-des-Arts;

SAINT-PÉTERSBOURG,

KLOSTERMANN père et fils, Libraires.

M. DCCC. XI.

Fig. 8.9 "Traité élémentaire des machines" (M. Hachette 1811 [56])

His book was generally based on three others but as a modern evolution of them, namely The "Theatrum Instrumentorum et Machinarum" by Jacobus Bessonus in 1578, the "Le Diverse et Artificiose Machine" by Agostino Ramelli in 1588 and

Jacob Leupold's books of 1724, "Theatrum Machinarum Generale" and "Teatrum Machinarum Hidraulicarum". Written for the machine course at the Paris École Polytechnique, this book was not limited to mechanism descriptions but it provided different machine calculations. For example, it included tables that listed the amount of water consumption in 1 day by a hydraulic machine with a specific piston diameter and height of water, as well as a theory of gears was described and detailed with design formulations.

Figure 8.10 shows the title page of the book by Betancourt and Lanz in which the programme for a mechanism course is reported as designed by Gaspar Monge.

Since then, for a period of several decades, several authors approached the systematic study of machines, such as "Traité complet de mécanique appliquée aux arts" by J. A. Borgnis in 1818–1823, "Principles of Mechanism" by R. Willis in 1841, and "Principles of the Mechanics of Machinery and Engineering" by Weisbach in 1848. All these works were technical machine compendiums with theoretical explanations of how the mechanisms and machines work. They differ from previous treatises such as the "Theatrum Machinarum", whose aims were mainly to form general viewpoints.

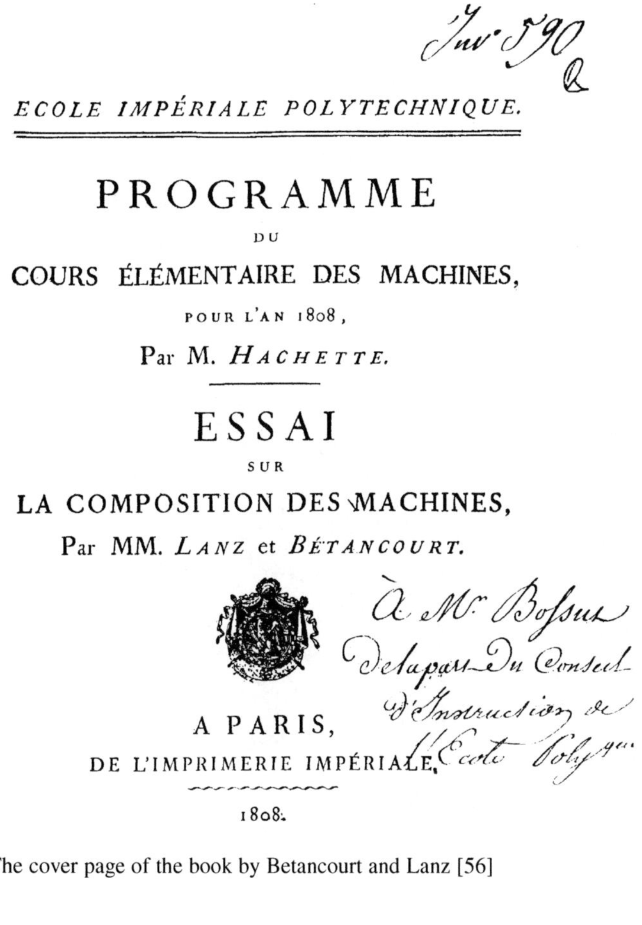

ECOLE IMPÉRIALE POLYTECHNIQUE.

PROGRAMME

DU

COURS ÉLÉMENTAIRE DES MACHINES,

POUR L'AN 1808,

Par M. *HACHETTE.*

ESSAI

SUR

LA COMPOSITION DES MACHINES,

Par MM. *LANZ* et *BÉTANCOURT.*

A PARIS,

DE L'IMPRIMERIE IMPÉRIALE.

1808.

Fig. 8.10 The cover page of the book by Betancourt and Lanz [56]

Gian Antonio Borgnis's encyclopaedic work is worth mentioning in detail. He was trained at the École Polytechnique and was a teacher in Pavía (which shows the extent to which TMM had become international). His work was recognised as a technical handbook and was used throughout the nineteenth century as a reference manual for professionals and researchers working on machines and mechanisms.

Borgnis classified machines into six types: receptors, communicators, modificators, frames, regulators, and operators. But this classification was not accepted by his contemporaries who continued to be in favour of Betancourt and Lanz's classification. Figure 8.11 shows an example from his work.

The difference between Borgnis' drawings and those by Willis (Fig. 8.12) and by Weisbach's (Fig. 8.13), can be observed as they evolve from theoretical diagrams towards a representation with constructional details.

The evolution was not as drastic as between the Middle Ages and the Renaissance but it slowly evolved from an artistic representation of machines towards a technical-ruled content.

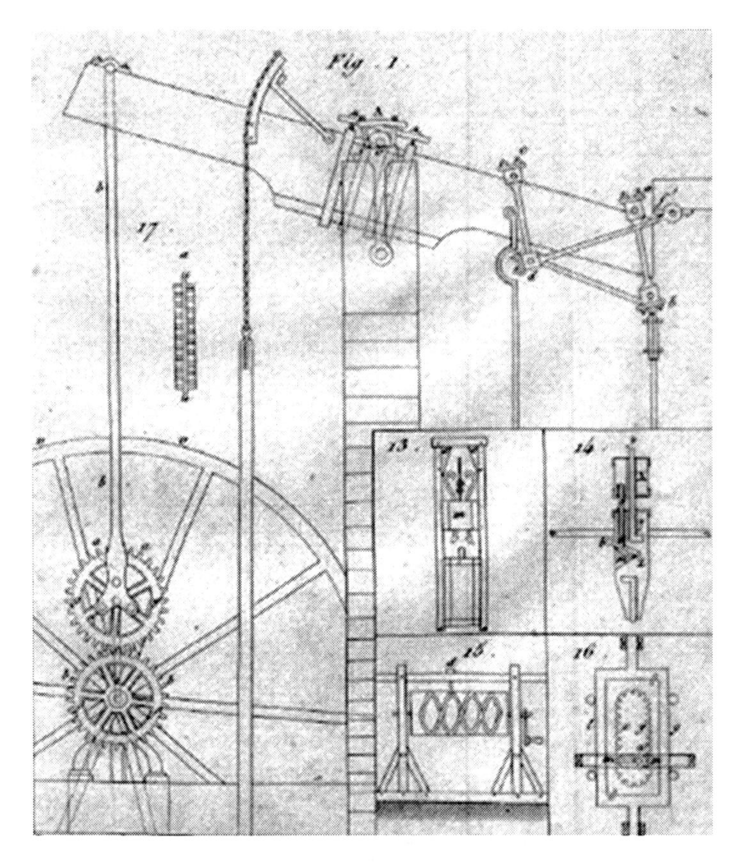

Fig. 8.11 Machine and mechanism drawings from the "Traité complet de mécanique appliquée aux arts" by J. A. Borgnis [22]

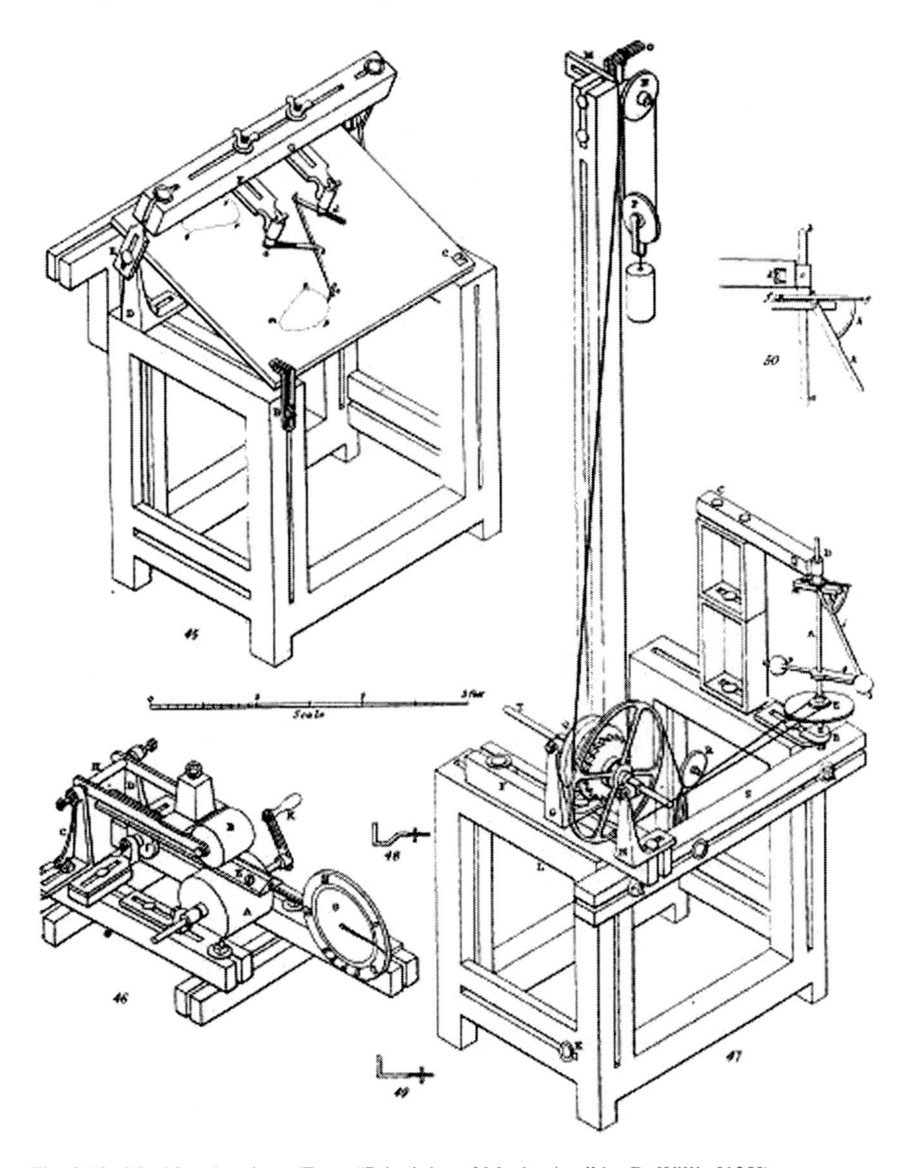

Fig. 8.12 Machine drawings (From "Principles of Mechanism" by R. Willis [139])

These books on the theory of mechanisms contain all kinds of machines as they are based on mechanisms that have been studied. In general, they deal with the principles of the examined mechanisms, but not their purposes or practical applications. Figure 8.14 illustrates some title pages to show the variety of these books, the extent to which TMM was disseminated worldwide, and its continuity throughout the nineteenth century.

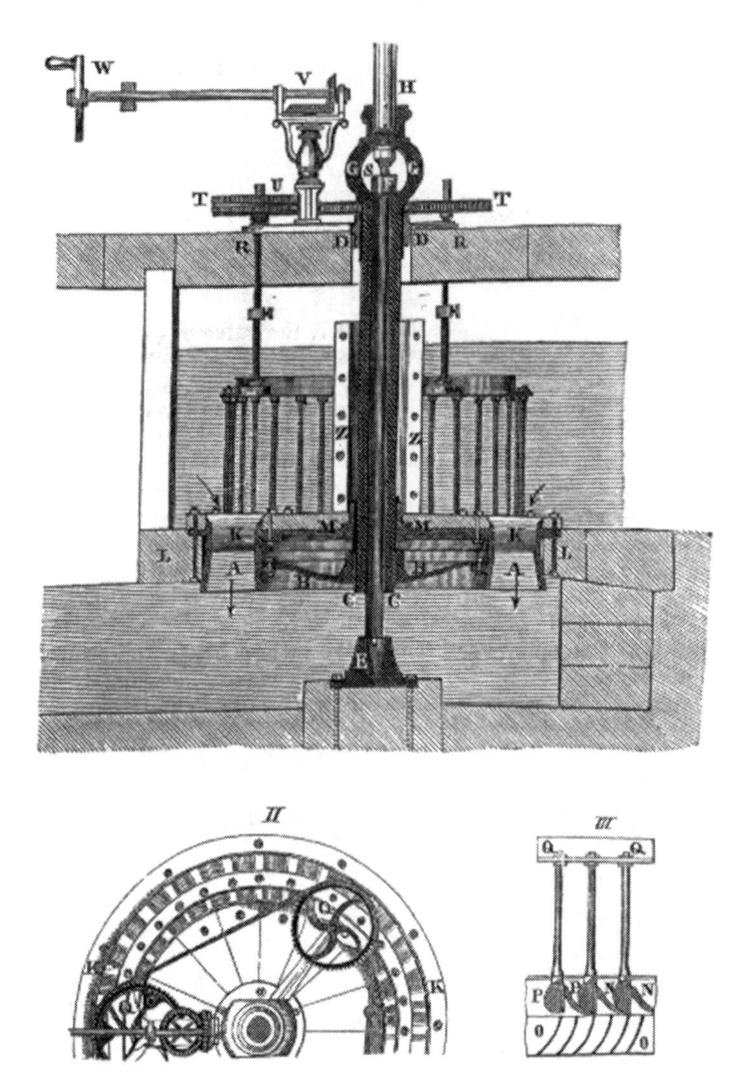

Fig. 8.13 Illustration from "Principles of the Mechanics of Machinery and Engineering" by J. Weisbach [138]

Labouyale in his "Traité de cinématique" and "Théorie des mécanismes" (1861) and Redtenbacher in his "Die Bewegungs-Mechanismen" (1866) also summarised the types and scope of each mechanism, although Labouyale chose to maintain Betancourt and Lanz's classification, while Redtenbacher chose to classify machines according to their use.

Figure 8.15 shows an illustration from Labouyale's book and Figs. 8.16 and 8.17 show illustrations from Redtenbacher's textbook.

a

RAPPORT
FAIT A LA PREMIÈRE CLASSE
DE L'INSTITUT,
LE SEIZE JUIN 1806,
PAR MM. MONGE, COULOMB ET PRONY,
SUR LE PROJET
D'UNE NOUVELLE MACHINE HYDRAULIQUE,
POUR REMPLACER
L'ANCIENNE MACHINE DE MARLY,
Présenté par Mr. BAADER.

A PARIS.
M. DCCC. VII
DE L'IMPRIMERIE DE MARGUERITE CHOLLET.

b

TRAITÉ COMPLET
DE MÉCANIQUE
APPLIQUÉE AUX ARTS,
CONTENANT l'Exposition méthodique des théories et des expériences les plus utiles pour diriger le choix, l'invention, la construction et l'emploi de toutes les espèces de machines ;

PAR M. J.-A. BORGNIS,
INGÉNIEUR ET MEMBRE DE PLUSIEURS ACADÉMIES.

Composition des Machines.

PARIS,
BACHELIER, LIBRAIRE, QUAI DES AUGUSTINS.
1818.

c

DIE
BEWEGUNGS-MECHANISMEN
Darstellung und Beschreibung eines Theiles der
MASCHINEN-MODELL-SAMMLUNG
der polytechnischen Schule in Carlsruhe.
von
F. REDTENBACHER
Großherzoglich Badischer Hofrath u. Director an der polytechnischen Schule in Carlsruhe.

Neue Auflage.
Mit 80 lithographirten Tafeln.

HEIDELBERG
VERLAGSBUCHHANDLUNG VON FRIEDRICH BASSERMANN

Fig. 8.14 Title pages of (**a**) a class by G. Monge, 1807; (**b**) book "Traité complet de mécanique" by Borgnis, 1818 [22]; (**c**) book "Die Bewegungs-Mechanismen" by Redtenbacher, 1866 [95]

In 1871, Brown published "Five Hundred and Seven Mechanical Movements" (Fig. 8.18). Using the 507 drawings referred to in the title, he described all the types of mechanical movements that were known at the time.

Very shortly after, Franz Reuleaux published six books on machinery, the most important of which were "Kinematics of Machinery" in 1875 and "Lehrbuch der Kinematik, V.1 – Theoretische Kinematik" in 1876. In those books he discussed the

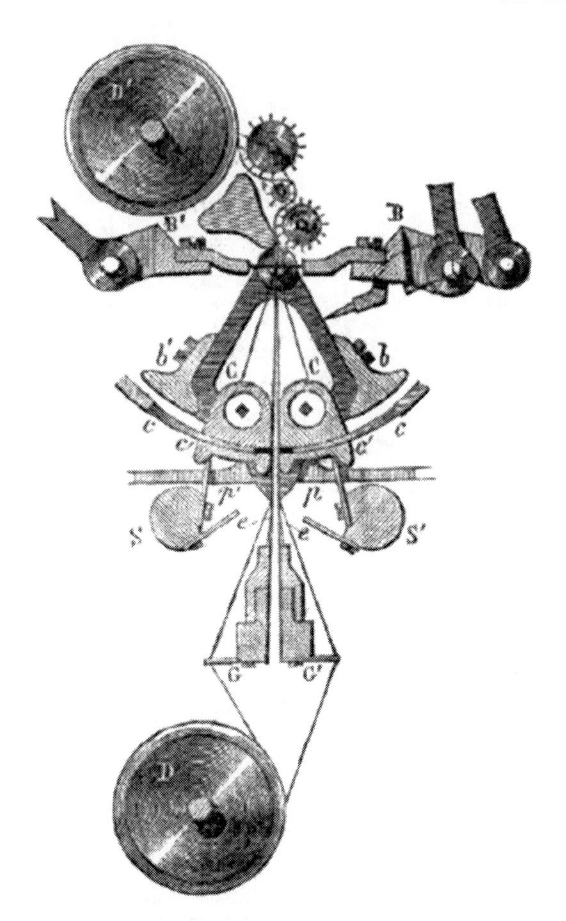

Fig. 8.15 Mechanism (From Labouyale's "Traité de cinématique" (1861) [69])

origins of machines and topics such as how to improve kinematic principles. He also stated the need for a specific terminology to define kinematic concepts, as well as the importance of studying machines based on an analysis of the movements of the component parts.

Figure 8.19 reproduces the title page from "The Kinematics of Machinery" by F. Reuleaux (1975) translated by Alex B.W. Kennedy in 1876.

The books by Reuleaux were a modern evolution of the works and approaches by Ramelli, Weisbach, Labouyale, and particularly by his teacher Redtenbacher. They were also based on a study of new and recently introduced machines, like the steam engine.

Reuleaux divided the elementary elements of machines into 21 categories, among which he considered screws, couplings, flywheels, chains, and springs. He also proposed a subcategory for any kind of machine as a main mechanism, a distribution (split into feed and abduction), a regulating mechanism (including braking mechanism), and

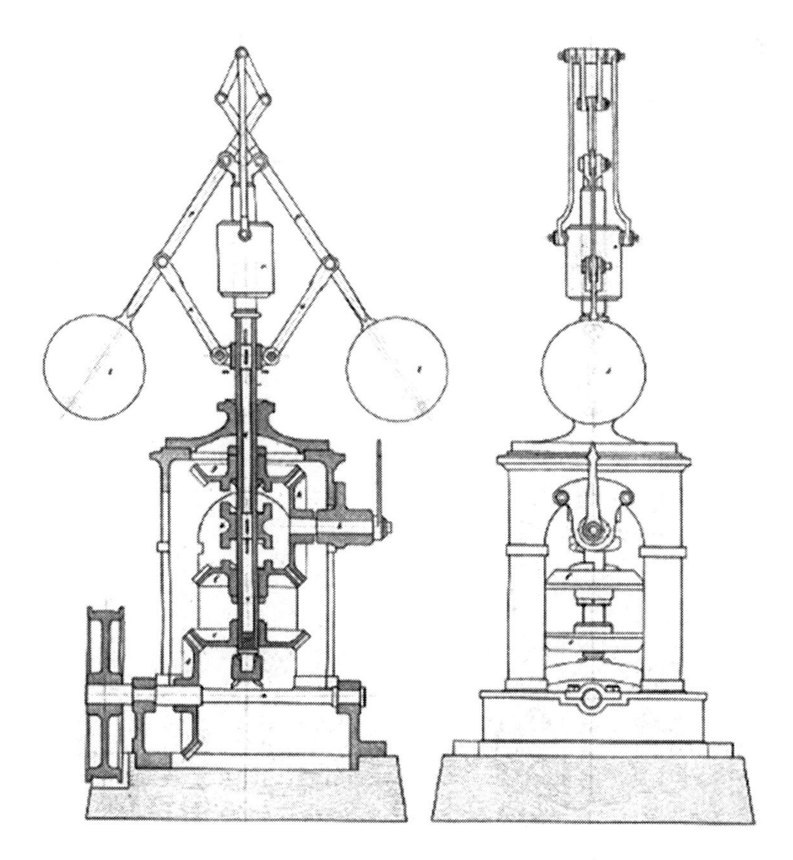

Fig. 8.16 Speed regulator (From "Die Bewegungs-Mechanismen", F. Redtenbacher (1866) [95])

a transmission. One of Franz Reuleaux's contributions is also the clear system of symbols that can be used to identify kinematic parameters as are still in used today.

Following in the footsteps of Reuleaux, Schröder in his "Catalog of Reuleaux Models" (1899) and Voight in "Kinematische Modelle nach Prof Reuleaux" (1907) confirmed the teaching approach of his predecessor with great skill. The aforementioned steam engine is an example of this. It is shown in Fig. 8.20 which is an illustration taken from Schröder's book and the illustrations from Reuleaux's classifications are shown in Figs. 8.21 and 8.22 are from Voight's book.

On Progress in Practical Use

Simultaneously to academic machine study, design issues attracted great interest in the nineteenth century to actually implement both these studies and practical results in several fields of industrial engineering. This led to considerable improvements in

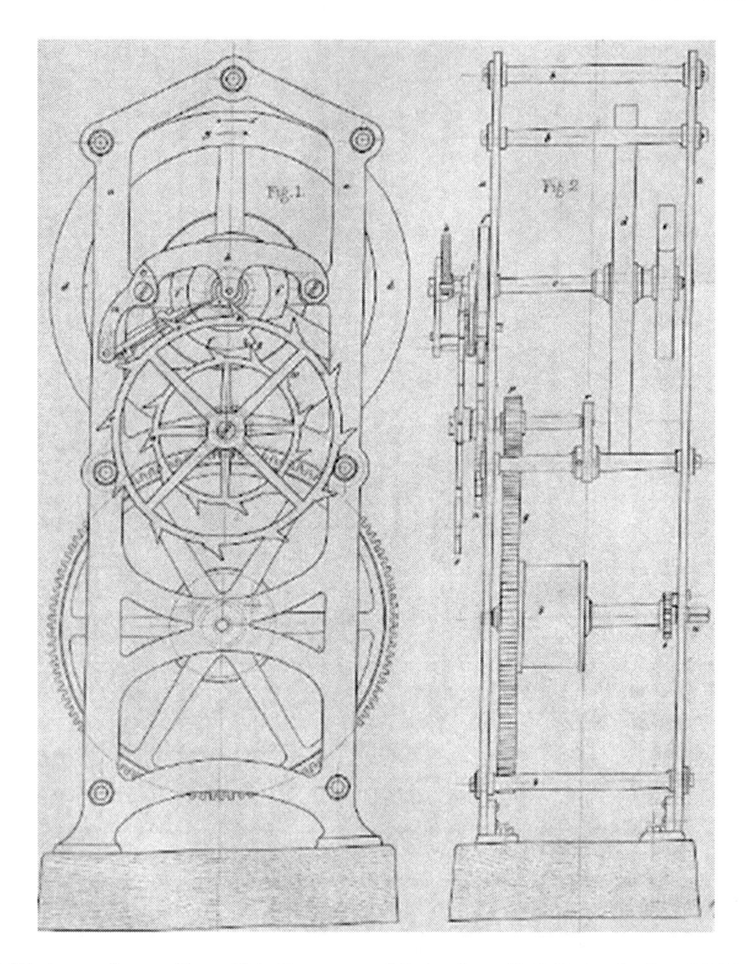

Fig. 8.17 A set of gears (From "Die Bewegungs-Mechanismen", F. Redtenbacher (1866) [95])

the automation and performance of production processes. Examples of those applications and new design are reported in Fig. 8.23a as an automatic loom, in Fig. 8.23b as an automatic saw, in Fig. 8.24a as the power transmission mechanism for the first railway engines, and in Fig. 8.24b as a machine for calculating and writing.

The previous chapter has stressed the Industrial Revolution's role in promoting the development of machines and mechanisms. But the great technical leap forward began in the twentieth century when many of the successes achieved in the progress of mechanical engineering could be ascribed to the intense teaching of theory and practice that was developed in the second half of the nineteenth century. This can be emphasised as documented in the afore-mentioned texts, as well as in many texts, as a collective memory of numerous engineers and scientists who developed machinery and mechanisms both in theory and practice.

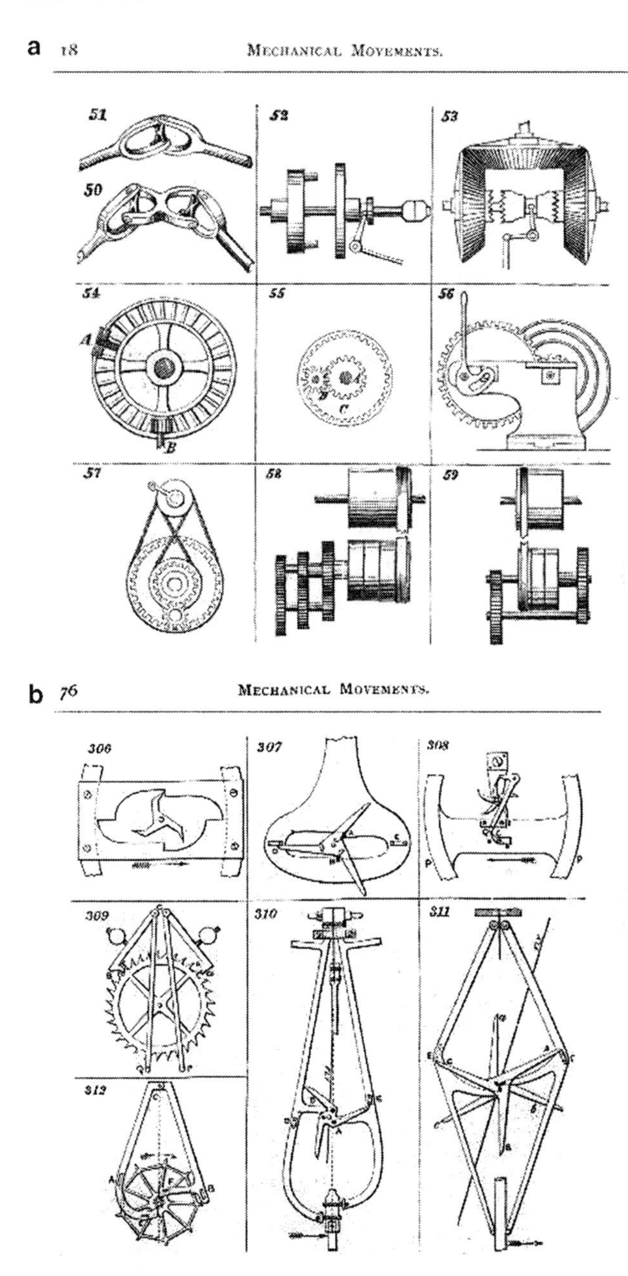

Fig. 8.18 Drawings of mechanisms (From "Five Hundred and Seven Mechanical Movements", H. T. Brown (1871) [27])

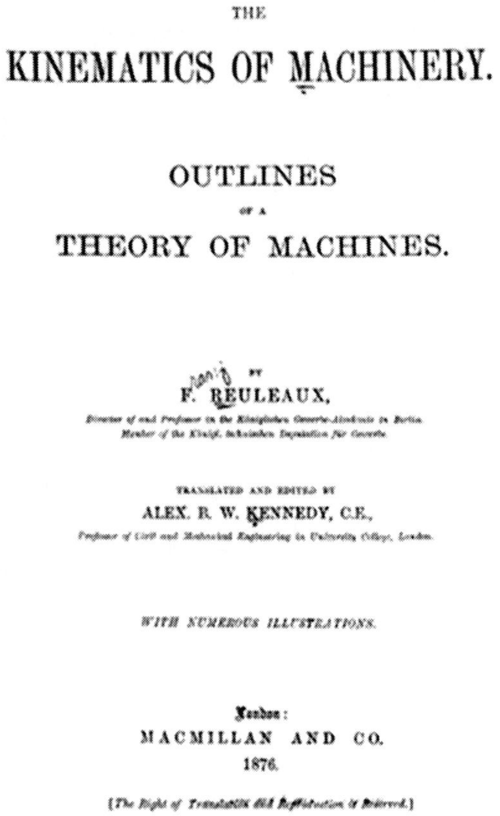

THE

KINEMATICS OF MACHINERY.

OUTLINES

OF A

THEORY OF MACHINES.

BY

F. REULEAUX,

Director of and Professor in the Königlichen Gewerbe-Akademie in Berlin.
Member of the Königl. technischen Deputation für Gewerbe.

TRANSLATED AND EDITED BY

ALEX. B. W. KENNEDY, C.E.,

Professor of Civil and Mechanical Engineering in University College, London.

WITH NUMEROUS ILLUSTRATIONS.

London:

MACMILLAN AND CO.

1876.

[The Right of Translation and Reproduction is Reserved.]

Fig. 8.19 First translation of the Kinematics of Machinery by Alex B.W. Kennedy (1876) [66]

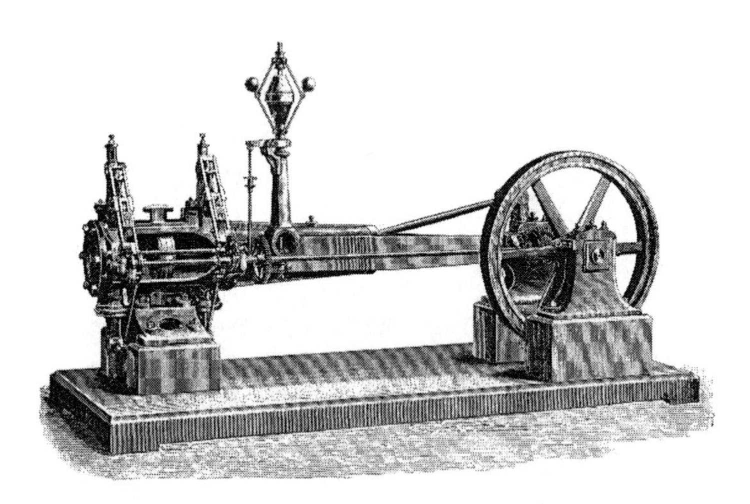

Fig. 8.20 Steam engine (From the "Catalogue of Reuleaux Models". Schröder (1899) [110])

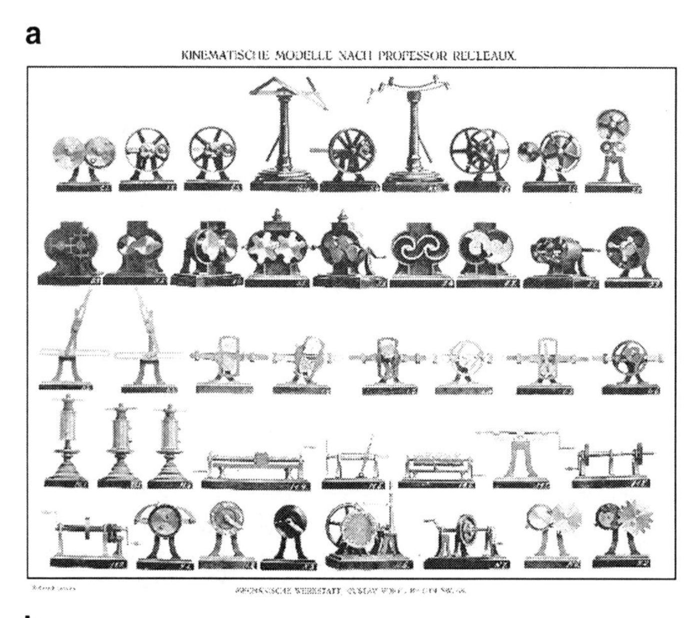

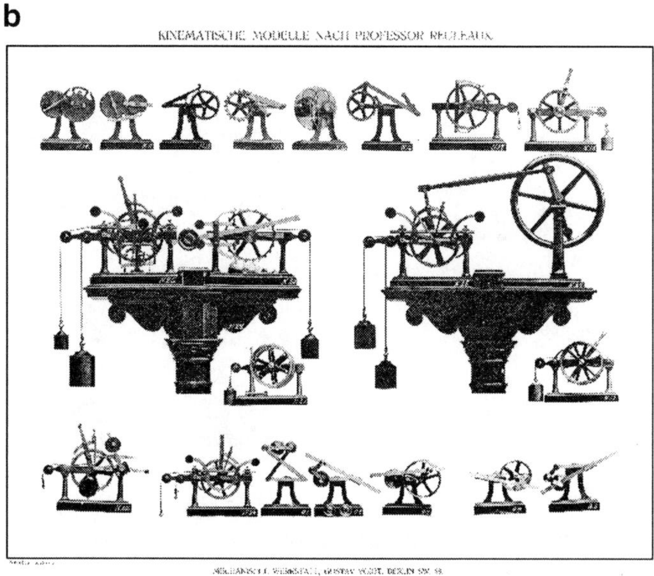

Fig. 8.21 Mechanisms (From "Kinematische Modelle nach Prof Reuleaux", Voight (1907) [136])

On Mathematization of Mechanism Design

The design of machinery and mechanisms evolved from being mainly based on almost pure geometry to more mathematical formulas. Lorenzo Allievi's work (1895) stands out as a brilliant example of this and it may be understood as a mathematical reformulation of a large part of Burmester's geometric theory (1888).

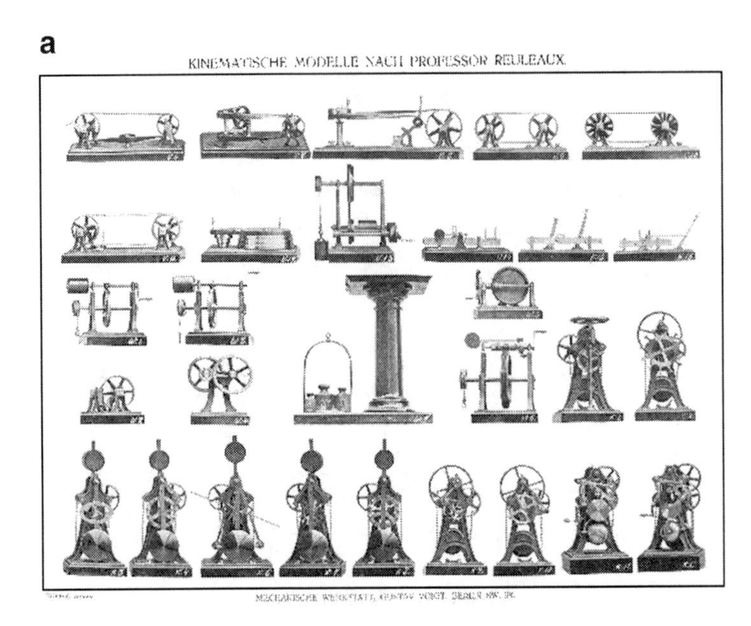

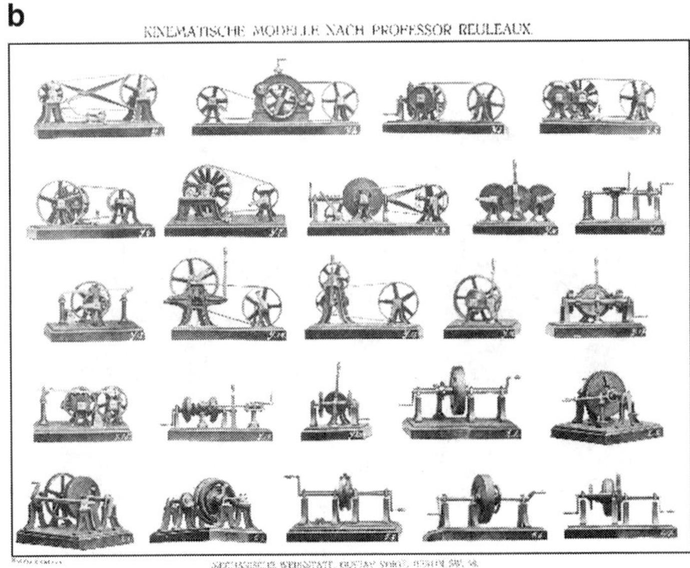

Fig. 8.22 Mechanisms (From "Kinematische Modelle nach Prof Reuleaux", Voight (1907) [136])

In addition, many authors successfully strove to update the theory of machines and mechanisms, by considering the acquired technological innovations and experience, in new analytical and design procedures. Examples of this fecundus activity were the works of Bourguignon (1906), Franke and Oldenbourg (1930), Bricard (1927), and Grubler (1917), to name but a few.

a

b

Fig. 8.23 (**a**) Textile machine and (**b**) automatic saw [32]

However, during the two world wars there was a noticeable lack of meaningful publications, probably because most technical know-how was kept secret for military reasons.

The beginning of the 1950s saw a renewed interest in the design of machines and mechanisms. This was most likely due to the optimistic perspectives that were provided by the new computer-aided analytical and calculating resources that had

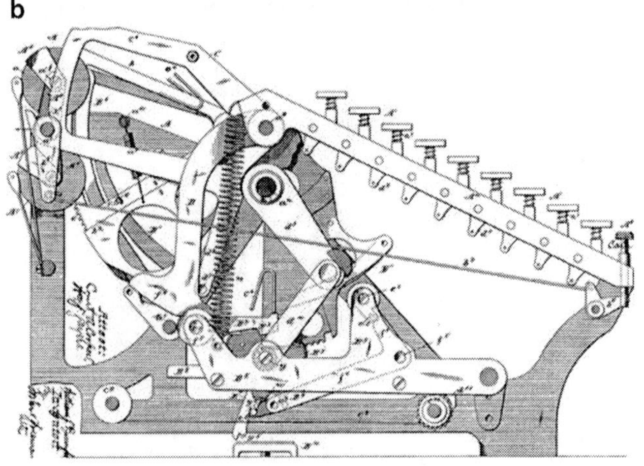

Fig. 8.24 (**a**) Power transmission of a railway engine, (**b**) calculating/writing machine [32]

just appeared with electronic systems. Thus, the kinematics of machinery was reviewed with the help of these new tools. In this respect, we can mention works by Rosenauer and Willis (1953), Beggs (1955), Pasillo (1961), Hain (1967), and Hartenberg and Denavit (1964). They may all be also thought of as representatives of the role that North America was beginning to play in the design of machines and mechanisms as a continuation of the European tradition.

Figure 8.25 refers to two early works (but are still valid) that represented the current trends in machinery computation. In Fig. 8.25a, by Freudenstein and Sandor (1959), the kinematics of a four-bar plane mechanism is modelled for an analytical model for automatic calculation. This computer-aided study was the first to use the computer for designing mechanisms. The model in Fig. 8.25b, by Roth and Freudenstein (1963), was developed to investigate the numerical aspects of computer-aided calculation as applied to the study of mechanisms.

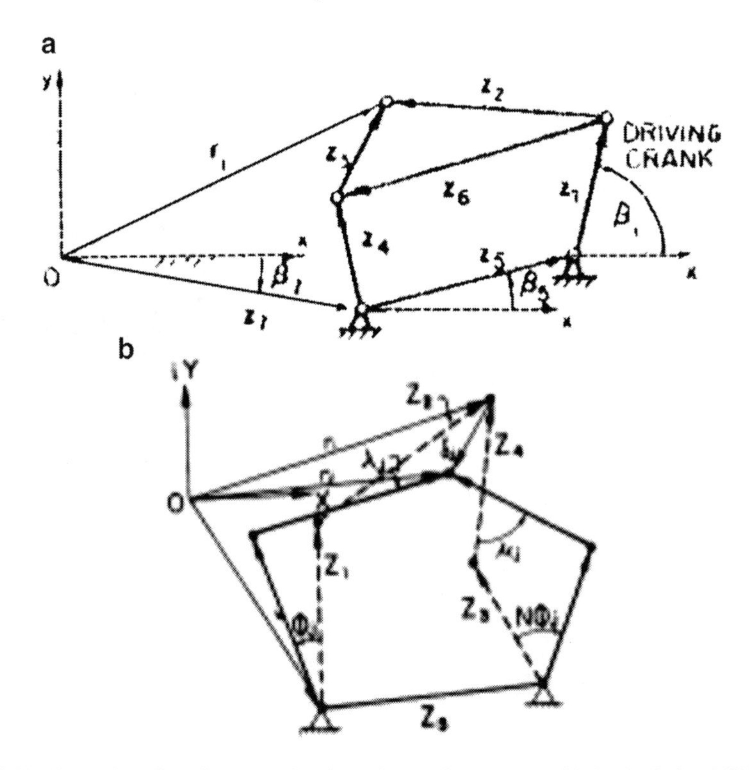

Fig. 8.25 Examples of the first models oriented towards computer-aided calculation [43]

Indeed, the computer has renewed interest in the formulation and design of mechanisms with new and old solutions coming together at the same time.

The enormous potential of modern means of calculation has provided a huge stimulus for approaching the study of complex mechanisms. This is the case, for example, of the parallel manipulators shown as a Gough-Stewart platform in Fig. 8.26a, Parallel manipulator architecture was first used in some very limited industrial applications, but during the last 2 decades has attracted great interest both from theoretical and practical viewpoints.

Because of augmented computation capabilities, ever more complex mechanisms began to be studied, sometimes only for purely theoretical interest, as is the case with Fig. 8.26b. This shows an exercise for applying numerical methods without any specific practical application in mind.

Current research activities and applications in the design of machines and mechanisms may be considered to be mainly computer-based and therefore mainly oriented towards developing procedures for analysis and synthesis rather than seeking new designs. Due to the enormous number of researchers working in this field and the many ways of dealing with mechanism-related problems, it is impossible to mention all the recent lines of research.

One example of the use of computers as a tool for studying mechanisms, both in the academic field and in the industrial sectors, can be considered to be the early

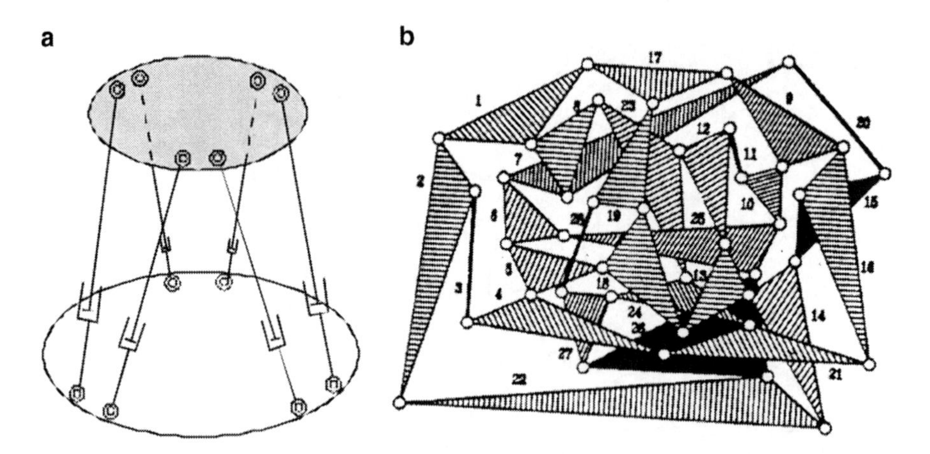

Fig. 8.26 Examples of complex mechanisms studied in the 1960s: (**a**) parallel architecture or the Gough-Stewart platform; (**b**) plane mechanism with 28 bonds and one degree of freedom [32]

experience with the "European Systems Engineering Symposium" that was organised by IBM in Barcelona in 1965 (see Fig. 8.27 showing one of the illustrations from the paper entitled "A computer-oriented method for calculus of mechanisms" presented in this Symposium by Professor Emilio Bautista Paz).

However, debates arising in recent years in forums such as IFToMM (International Federation for the Promotion of Mechanism and Machine Science) congresses or the ASME (American Society of Mechanical Engineers) may show what have been and what are the major topics of interest in recent years in the science of mechanisms.

On Machine Training

Since the modern reformulation of machine kinematics in the 1950s, there have been many initiatives to establish suitable study plans for training engineers in this subject.

In the 1950s, studies in the theory of machines and mechanisms became widespread with courses in kinematics, dynamics, and the design of machines and mechanisms given in almost all universities around the world. Many texts have been written in many languages, which means that a huge bibliography has been developed on the subject.

As an example of Spain's specific contribution to the publications at that time, two texts can be cited that are still valid and in current use. The book "Machine Kinematics and Dynamics", by Adelardo de Lamadrid and Antonio de Corral (1963) is an update of the original that was published by Adelardo Martínez de La Madrid in 1948 (Fig. 8.28 shows a drawing from the book). The book "Synthesis

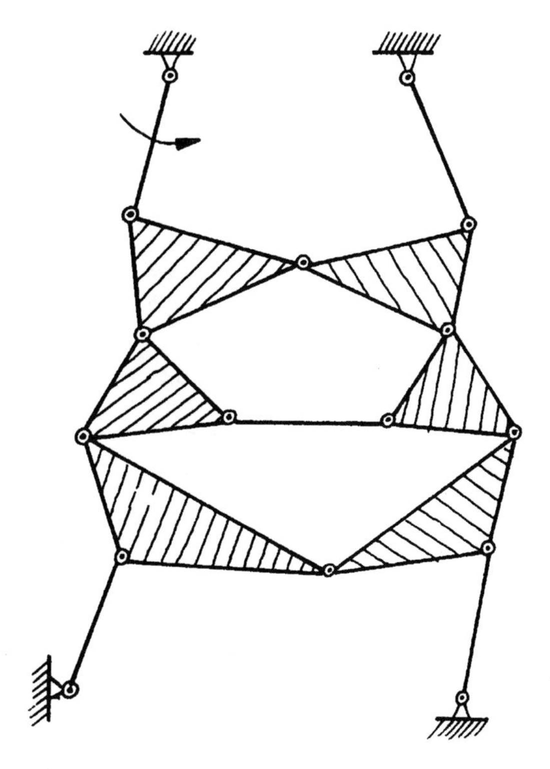

Fig. 8.27 Illustration from "A computer-oriented method for calculus of mechanisms" presented at the "European Systems Engineering Symposium" by Professor Emilio Bautista Paz [17]

of Mechanisms" by Justo Nieto (1978) (Fig. 8.29 shows an illustration from the book) is an interesting work for the systematization of kinematic procedures for mechanism synthesis with a modern approach.

The academic atmosphere of recent decades has been dominated by texts from the English-speaking world. A large part of machine teaching in western countries or those technologically belonging to their area of influence, has been based on publications in English. Examples are the books: "Design in Mechanical Engineering" by J.E. Shigley and L.D. Mitchell; "Machine Mechanics" by C.W. Haw, E.J. Crane and W.L. Rogers, "Machine Elements" by K.H. Decaer, and "Principles of Tribology" by Halling, to cite just a few examples.

For obvious reasons, in Eastern Europe a similar bibliography on machines was produced through Russian technological domination. The political vocation of Russia to spread its zone of influence to the entire world gave rise to a strange situation from the bibliographical aspects: the State itself supported the translation of texts into other languages, especially through Mir Publishing ("Mir" means Peace in Russian). This meant that Russian texts on machines were extensively used by universities in other countries.

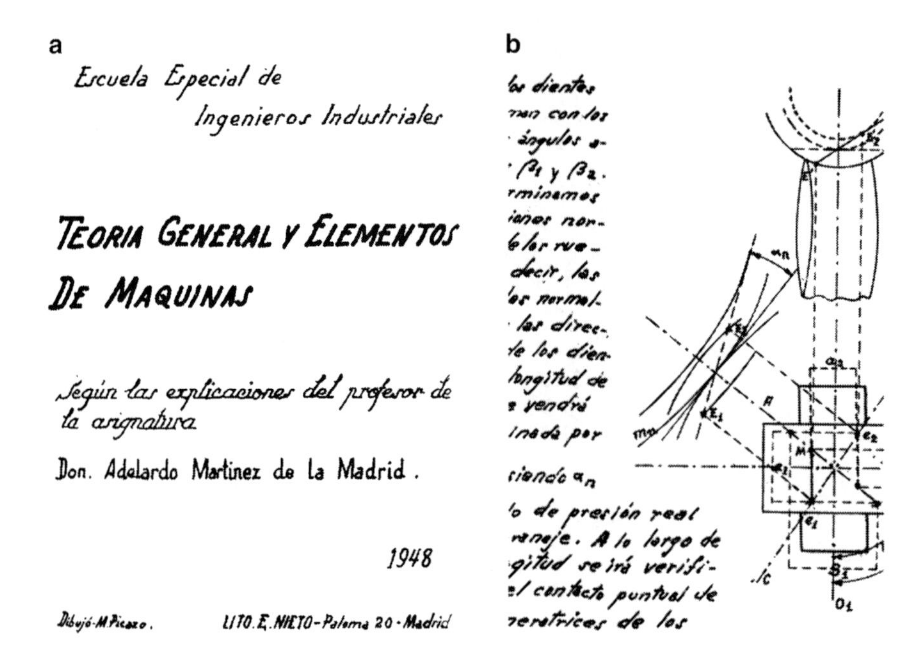
a

Escuela Especial de
Ingenieros Industriales

TEORIA GENERAL Y ELEMENTOS
DE MAQUINAS

Según las explicaciones del profesor de
la asignatura

Don. Adelardo Martinez de la Madrid.

1948

Dibujó-M.Picazo. LITO.E.NIETO-Paloma 20-Madrid

b

Fig. 8.28 "Machine Kinematics and Dynamics" by A. de Lamadrid and A. de Corral (1948) [70]: (**a**) front cover; (**b**) a page

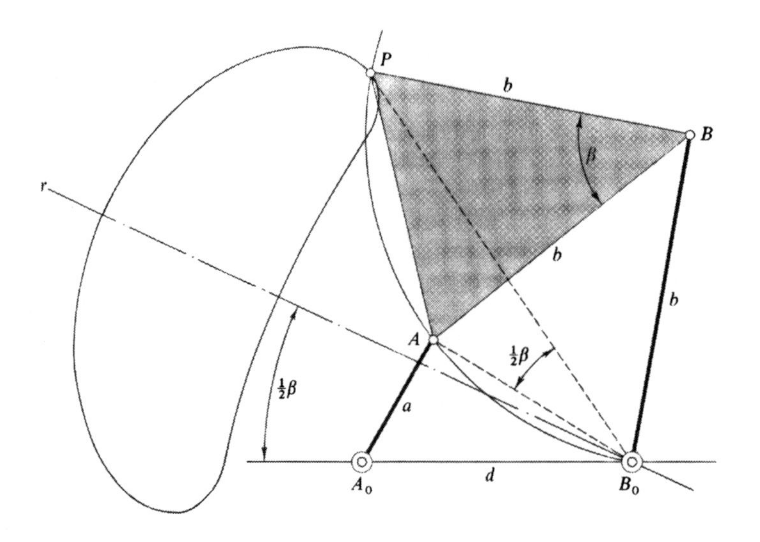

Fig. 8.29 An illustration from the text "Synthesis of Mechanisms" by Justo Nieto (1978) [89]

To quote just one example, we must mention the texts "Mechanisms in Modern Technology" by Ivan Artobolevski that were published in Moscow by Mir Publishers in 1976–1981 as an exhaustive compilation of mechanisms that may be considered as a complete machine encyclopaedia. The books contain descriptions of about 1,700 mechanisms with details of their operation capabilities and design peculiarities. The aim of these books is an exhaustive classification of mechanisms. Even today, this may be considered as one of the most thorough classifications. Figure 8.30, shows the gearwheel lever mechanism of a centrifugal tachometer, which is an example of the widespread circulation of Artobolevski's works in other languages that Russian.

Compared to the above-mentioned examples, books on machines in other languages (French, German, Italian, Japanese, etc.) have been less widely circulated and read and thus may have had less influence on teaching and research activities.

244	MECANISMO DE PALANCAS CON RUEDAS DENTADAS DE UN TACÓMETRO CENTRÍFUGO	PRD ME

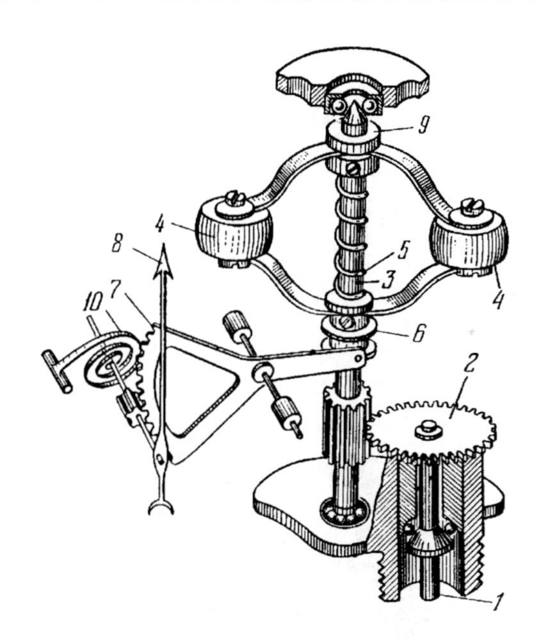

La rotación del árbol de distribución se transmite al árbol *1* del tacómetro y, por intermedio de la rueda dentada *2*, al eje *3*. Al girar el eje *3*, las cargas *4*, bajo la acción de la fuerza centrífuga, separan comprimiendo el resorte *5*. El desplazamiento del manguito *6* sobre el eje *3* del instrumento provoca el giro del sector dentado *7* y de la aguja *8* que registra la velocidad de rotación. El manguito *9* está rígidamente fijado sobre el eje *3*. La espiral *10* está destinada para eliminar los juegos en el mecanismo.

Fig. 8.30 A mechanism study from "Mechanisms in Modern Technology" by I. Artobolevski published in Moscow by Mir Publishing in 1976 [12]

Indeed, there is a trend for any publication to become a satellite of a publication in English, in the sense that interest is shown by authors to have an English edition.

But this less influential role of other language publications in no way means that machine publications have declined. One significant example, also because of its encyclopaedic nature, is "Machine Elements" by G. Niemann (1981). Figure 8.31 shows an illustration from the first edition in the original German language.

This book, that has been translated into many languages (see an illustration from the Spanish edition of "Machine Elements" by G. Niemann, in 1987 (Fig. 8.32), is an example of how the training tradition still prevails in the machine field of the great European schools.

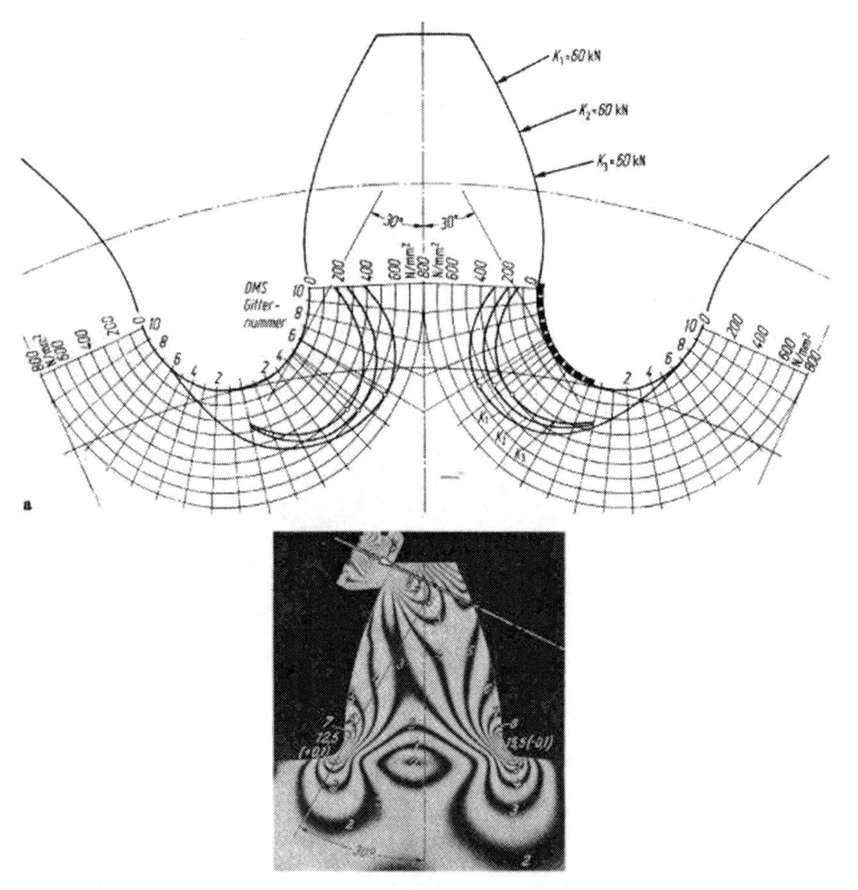

Bild 21.7/7. Ermittlung der örtlichen Zahnfußspannung. a) Mit Dehnmeßketten [21.8/97, 99]; b) mittels Spannungsoptik; ferner eignet sich die Methode der finiten Elemente oder der Integralgleichungen.

Fig. 8.31 Illustration from the original German edition of "Machine Elements" by G. Niemann (1981) [88]

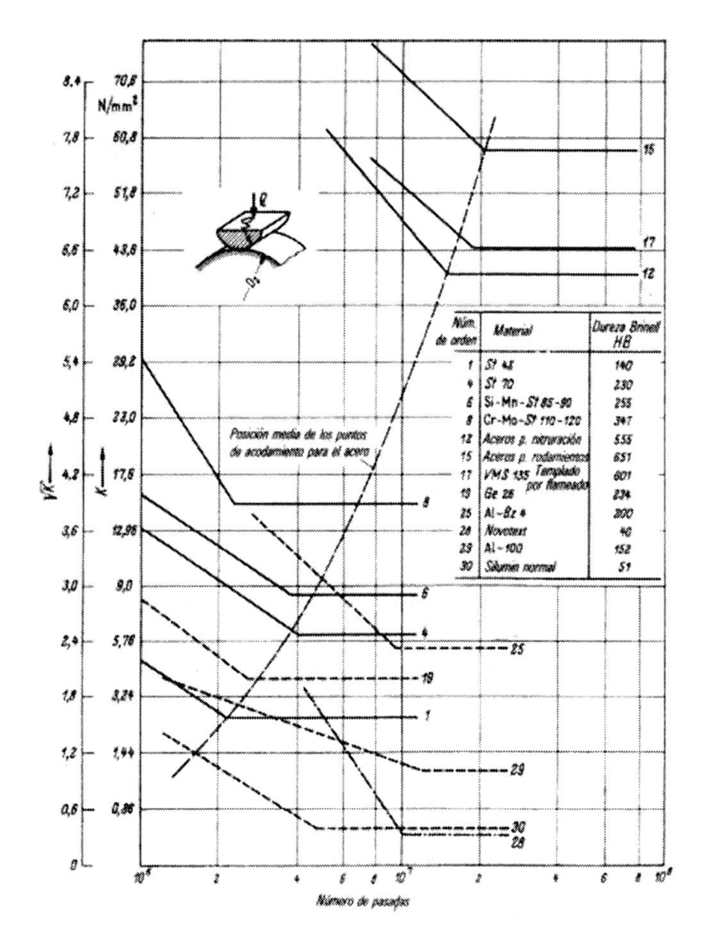

Fig. 8.32 Illustration from the Spanish edition of "Machine Elements" by G. Niemann (1987) [88]

Final Remarks

As we stated at the beginning, this book was planned with a complementary perspective on the general history of mankind from the point of view of the underlying mechanical ingenuity that has always been linked to social evolution. Since long past, human development has been conditioned by machine inventions which have also marked important historical events such as wars, trade expansions, and periods of cultural supremacy.

The reader could have the impression that the book stops before getting to the present-day period that is very fertile in the development of machinery. But there are more than sufficient reasons for this, in the opinion of the authors, and not only because the current final period is extremely short as compared with the examined period of time. More important is the difficulty in having a perspective on the historical significance and importance of what is contemporary. For example, it would have been difficult for a citizen of Athens to have been aware of how important the birth of philosophy was for the future developments of society and science, likewise for a Florentine to perceive the impact of the Renaissance in modern assessments.

This is why this book deliberately ends with an analysis of the considerations on machine repercussions with a perspective. The dissemination promotion through printed publications regarding machines and mechanisms, the inclusion of these publications as a means for higher studies, and systemising and rationalising mechanism analysis and design, have all contributed to the evolution to levels that have subsequently not been surpassed. Therefore, it can be considered a logical decision to consider perspective remarks as the end of a history of machines.

Looking at the Future

Although it was defined only 2 decades ago, Mechatronics represents an important direction for the future of mechanism science. Mechatronics is the integration of many aspects of mechanics with electronics, electricity, computing, and other things, since modern systems are made of many elements whose design and function are related to a variety of engineering disciplines. However, it should be emphasised that mechanical engineering will continue to play an important and essential role, since the mechanical nature of Man's actions for whom systems, modern and mechatronic, will be developed with an increasing number of "no longer necessary" mechanical components.

Likewise, the creation of mechanical microsystems is opening a wide field of possibilities. Photographic manufacturing methods similar to those that are used for electronic microcircuits we predict, will be used for the industrial development of microscopically sized machines in the future. Atomic and molecular structures may even be used as machine parts, as in the example shown in Fig. 8.33.

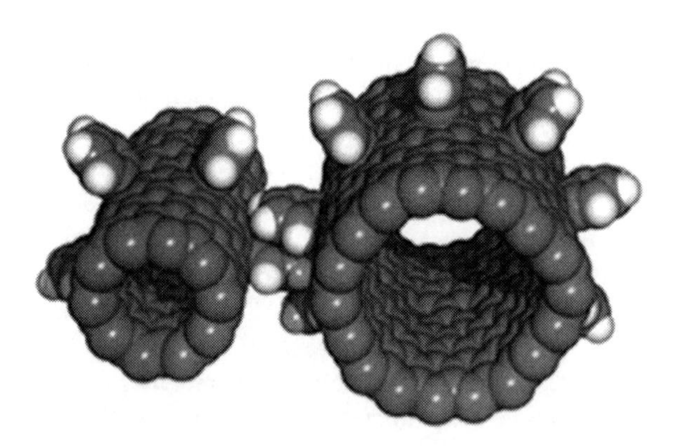

Fig. 8.33 Silicon nanogears

In addition, a current relevant fact that may mark the future of mechanical design is the recent interest in the study of biomechanics. This has spread in a few decades to numerous fields of industrial application. The design of many products, ranging from training shoes and furniture to vehicles and space capsules, takes careful account of biomechanics.

Many universities and companies have groups working in biomechanics, often collaborating with teams of doctors, since in many cases the attached problem or product will be for direct use in the human body.

Broadly speaking, the applications of biomechanics can be classified under three headings depending on what their aims are:

- To predict how the human body will behave when faced with external mechanical actions.
- To artificially strengthen the human body in its mechanical performance.
- To replace parts of the human body to ensure mechanical efficiency.

There are obviously works that are not oriented towards the human body and would therefore remain outside this classification. However, the social and scientific importance of these additional aspects is more restricted, although they may be of great scientific interest.

Biomechanics also serves as a support for developing clinical and surgical equipment although in a way that is typical of any mechanical engineering application. An artificial kidney or an oxygen economiser are examples of those possibilities.

Maybe the most spectacular developments of recent years are related to bone prostheses: knees, hips, teeth, legs, or entire arms that have either suffered irreversible damage due to injury or progressive deterioration due to age and which can now be replaced by industrial products, thereby considerably improving the quality of human life.

Any other type of prosthesis for the circulatory or urinal system, etc, always has some mechanical component.

Linked to some extent to the above-mentioned fields is the area of rehabilitation, where biomechanical studies have enabled us to regain the operation of parts of the body that are damaged by causes such as injuries and age.

The economic interest by insurance companies has resulted in an important role for this application of biomechanics. Rehabilitation is cheaper than a disability pension and is also beneficial for the quality of human life.

Sports training uses techniques that are similar to those for rehabilitation. In this sphere, biomechanics can be used to improve athletes' performances. The worldwide increase in competitions has provided a strong boost for this application, as well as for the development of material and equipment related to sport as a mass phenomenon.

The biomechanical study of the human body enables its behaviour to be predicted when subject to external mechanical actions. These are usually collisions and vibrations in addition to the mechanical actions the human body can carry out in the environment surrounding it.

Safety requirements in working environments and for the users of industrial products have also provided a great boost for biomechanical studies. A spectacular application in this field is the study of vehicle accidents.

If safety is a requirement for the current legislation, comfort and ergonomics are factors that affect the quality that is perceived by an operator or user of an industrial product and therefore it will affect its market success. The search for these characteristics has also meant a fresh impetus for biomechanical studies.

All these achievements are important and undoubtedly they will mark the trends in mechanical innovation that are different from those of past eras. However, more significant is the change of attitude that this can create worldwide in mechanical engineering.

On the Challenge of Biodevices

A detailed look at the living organisms around us, at least through the eyes of an engineer, shows us a multitude of mechanisms that appear to have been carefully, knowingly, and meticulously designed for overcoming numerous constructional problems with advanced technology.

Up to now, Nature had always been a field of observation for scientists who strove to solve its laws. But it has now become a field of observation for engineers who, in the light of bioengineering, are discovering that infinite technical solutions can be had that have been confirmed by thousands of experiments, sometimes over millions of years.

Apart from being a salutary lesson for our design activity, trying to adapt Nature's technology, to develop its basic ideas, and to improve it for specific applications, can provide considerable reinforcement for all current technologies through the design challenge that each biodevice can provide.

In addition, examining biodevices would seem to tell us that there are no real limits to raw materials or energy sources that cannot be overcome by the appropriate knowledge of physics and resourceful technology- Production on an industrial scale of a leg or an engine that runs on grass may cease to be mere curiosities and will become something vital for mankind in the future.

Observing biomechanics from such a viewpoint suddenly opens up a breach in the limits of engineering and identifies a roadmap to a new industrial revolution, since it has been locked inside its own world for decades.

Nature is issuing us with an unavoidable challenge from every corner of our environment to make us aware that this multiple challenge to our technical and scientific skills can only result in a leap forward in engineering and research.

Thus, we come to the end of the cycle of this book that began with machines, before man, as mechanical designs without the intervention of human resourcefulness, examples of which can be found in Chapter 1.

Laboratory experiments have even been performed on robots that reproduce themselves. The mother and father generate children that inherit and share their genes, thus opening up the way to evolutionary designs like those of Nature. A recent article in Nature magazine describes this development conducted by Hod Lipson at Cornell University, and states, “Although the machines we have created are still simple compared with biological self-reproduction, they demonstrate that mechanical self-reproduction is possible and not unique to biology. It would be interesting to see if they spontaneously learn to reproduce using evolutionary principles”.

This copying attitude of engineered biodevices may become a new turning point in mechanical innovation.

To summarise, we can quote Frenay (“Darwin Among the Machines”, 2006) when he stated: “The culture currently shaping and surrounding us, was shaped in the machine age. Its success knows no precedents, but it is becoming ever clearer that machines – as we know them today – are merely a sub-component of biology, but in a primitive state, of course. The great leap forward in knowledge that created them and scattered them over the planet was nothing more or less than a first step out of the darkness of history”.

On the Challenges with Mechatronics

Mechatronics has been established since the 1990s as multidisciplinary engineering dealing with the complexity nature of modern systems. As shown in Fig. 8.34, mechatronics is understood as a fusion of the disciplines mechanics, electronics, electric engineering, control, measurement, and computer science. From IFToMM terminology, the mechanism basic role is emphasized with a definition of mechatronics as a ‘synergetic combination of mechanical engineering, electrical engineering, and information technology for the integrated design of intelligent technical systems, in particular mechanisms and machines’.

A mechatronic system is, indeed, composed of mechanical parts, electric devices, electronics components, sensors, hardware and it is operated and controlled by the supervisions and commands that are programmed through suitable software. Thus, the main characteristics of mechatronic systems are the integration and complementarities of several aspects from the many disciplines that describe

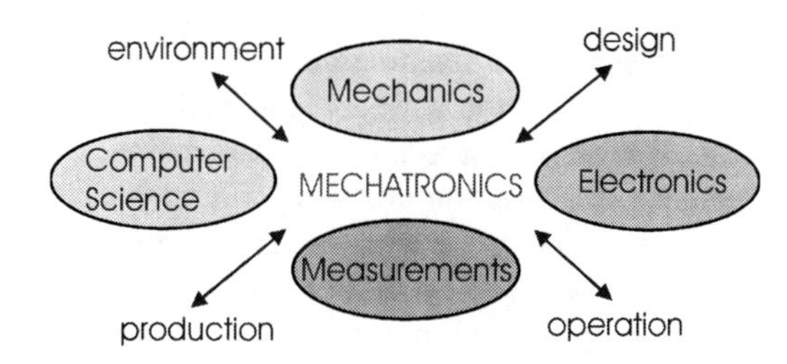

Fig. 8.34 A definition of Mechatronics

the design and operation of the components and overall system. In addition, for a suitable operation of a mechatronic system, engineering issues and human-system aspects must be considered as looking at features and constraints from the environment in which a system operates, the design by which it has been developed, the operation performance through which it fulfill the task, and the production by which it has been built.

Indeed, a certain mechatronic approach has always been considered in industrial engineering when the mechanisms and machines have been completed and integrated with other systems with the aim of enhancing and regulating the machinery operation in an efficient way and for user-oriented application. But it is to noted that the current level of integration of those different aspects in machinery has reached a complexity by which the machine operation strongly depends on the other electric and electronics systems that are controlled via informatics means that the mechanical core of a machinery is often hidden or even difficult to appreciate. However, that integration of mechanisms with other systems with different natures and operations has permitted the development of machinery to a very high level of efficiency and even a reduction of size and weight for the benefit of better operation and human-machine interaction.

The optimistic views on these achievements that indeed are due to high technology in mechatronics, very often bring several engineers (who have no solid grounding in mechanical engineering) to overestimate the importance of electric, electronic, and informatics systems over the mechanical core of a machine with its mechanical aspects, since these other systems are still evolving continuously and quickly with new solutions, while mechanical engineering seems to have reached a maturity of its full capability.

Nevertheless, today the awareness of the role of mechanical engineering and the importance of machines seem to be properly reconsidered and the challenges with mechanisms and machines will be considered in the future with great attention also with the aim of evolving mechanical aspects and related knowledge and techniques in order to have evolutions and enhancements similarly to those in the other disciplines in mechatronics.

Indeed, mechatronics is believed to be the future for industrial engineering, and even in other fields, since the specific formation of mechanical, electric, electronic engineering will be fused. Those mechatronic engineers will be able to design and operate the modern mechatronic systems.

Thus, the challenge of engineering of machines towards mechatronic engineering can be understood as a greater technical vision for systems that will become more and more complicated and integrated with a variety of multidisciplinary aspects. This amplification of interest and knowledge will be required to mechanical engineers and machine designers who, nevertheless, since time immemorial, have been presented with a vision of the systems as integrated with other components like power sources (in the past mainly from hydraulics and then from steam engines), transmission systems for motion or force depending on the application, and even the interface with human operators.

The evolution of professional engineers to mechatronic engineers is somehow already on-going through the reform of the teaching plans all around the world, with great efforts in revision, more that in the past, with a continuous update to include new knowledge of systems and new interdisciplinarity in the academic teaching. Similarly, the professional activity will require technical skills that will permit the engineers to understand not only the systems but even to understand each other or experts for the development of quality and quantity of new productions or services. Even the experts, who investigate in specific areas, will require a mechatronic culture not only to interact with the potential users of their research results but mainly to understand better the requirements and enhancements that are needed for solving problems and to develop new systems.

The historical evolution of the technical knowledge has required a differentiation of disciplines in the past with the growth of knowledge and its different aspects, so that the unique figure of the industrial engineer that was established at the beginning of nineteenth century, has evolved into a multitude of professional figures such as mechanical, electric, electronic, informatics, aeronautics, naval, hydraulics, civil, transport, etc., in the 1970s. But a further increase in the complexity and sophistication of the machines (now systems) today and more in the future will somehow require a reunification of the expertises in the unique figure of the mechatronic engineer.

However, whatever electronics, informatics, telecommunications, and so on, will be enhanced and expanded in mechatronics technology, mechanical design will always be needed, since man will always live and interact with the environment on the basis of mechanical phenomena of a human nature.

Chronic Table

TIME LINE	PREHISTORY							
Historical Event	Paeliolithic: 2.5 million years to 10000 BC.	Cro-Magnon man: approx. 35000 BC.	Mesolithic: 10000-7000 BC.	Göbekli Tepe Temples Analolia 9000 BC.	Neolithic: 7000 to 2000 BC.	Kamir Shahir, domesticated settlement: 6000 BC	Settlements of Jarmo and Hassuna: 4300 BC.	Uruk, first urban settlement: 3500 BC.
Scientific or Technological Event	Appearance of symbolic language: approx. 800000 BC.	First graphology: approx. 40000 BC.	Bone industry, body ornaments: approx. 10000 BC.	First stable farming communities (wheat and barley): 8000 BC.	Jericho culture, use of clay and adobe: 8000 BC.	Polished stone tools and fish hooks: 7000 BC	Pre-domestication of animals: Middle East 6500 BC	Megalithic culture: Carnac, Stonehenge 4000-2500 BC
2.- CHINESE MACHINES AND INVENTIONS								
3.- MECHANICAL ENGINEERING IN ANTIQUITY								
4.- MEDIEVAL MACHINES AND MECHANISMS								
5.- THE MACHINE RENAISSANCE								
6.- MACHINES OF THE IBERIAN COLONIAL EMPIRES								
7.- THE MACHINERY OF THE INDUSTRIAL REVOLUTION								
8.-SOME THOUGHTS ON MACHINES								

TIME LINE	ANTIQUITY (aprox. 3300 BC to 476 AC)										
Historical Event	Invention of Writing in Sumer: approx. 3300 BC.	Hieroglyphic Writing in Egypt: approx. 3100 BC.	Egyptian Civilisation: aprox. 3200 to 332 BC.	Phoenician Civilisation: approx. 3000 to 332 BC.	Babylonian Civilisation: 2003 to 539 BC.	Persian Civilisation : approx. 1000 to 330 BC.	Greek Civilisation: approx. 1400 BC. to 2nd century AD.		Roman Civilisation: 8th century BC. to 476 AD.		
Scientific or Technological Event	Beginnings of bronze metallurgy in Middle East: approx. 4000 BC	Building of the first Zigurats: 3000 BC	Building of the Great Pyramid of Kheops: 2589-2566 BC	First lunar calendar: Sumerians in 2300 BC	Bronze Age: aprox. 2000-700 BC in Europe	Iron Age: approx. 1300-1st century BC in Europe	Horse of Troy and taking of Troy by the Achaeans: 1183 BC	Aristotle (384 to 322 BC) Archimedes (287 to 212 BC)	Advances in Medicine and Surgery: Galen 129-199 BC	Hero of Alexandria (1st century BC-1st century AD.)	Vitruvius (1st century AD.)
2.- CHINESE MACHINES AND INVENTIONS											
	Domestic use of the silkworm and fabrics: 3000 BC	Knowledge of Astronomy: 2700 BC					"Kao Gong Ji" (770-221 BC.)	Iron metallurgy in China: approx. 500 BC	"Dictionary of Local Expressions" (15 BC)	Manufacture of paper in China: 115	Knowledge of magnetic stone: 120
3.- MECHANICAL ENGINEERING IN ANTIQUITY											
						Use of the precision balance in Egypt: approx. 1300 BC		Works of Archimedes (287-212 BC)	Works of Ctesibius (3rd-2nd century BC)	Hero's "Pneumática" (1st century BC-1st century AD)	"De Architectura" by Vitruvius (1st century AD)
4.- MEDIEVAL MACHINES AND MECHANISMS											
5.- THE MACHINE RENAISSANCE											
6.- MACHINES OF THE IBERIAN COLONIAL EMPIRES											
7.- THE MACHINERY OF THE INDUSTRIAL REVOLUTION											
8.-SOME THOUGHTS ON MACHINES											
							Machines and devices of the Greek and Roman civilisations, whose rediscovery in the 15th and 16th centuries would lead to the great scientific-technological advances of the Renaissance				

TIME LINE	MIDDLE AGES (476 - 1453)									
Historical Event	Fall of Roman Empire in the West: 476	The Hegira: Mohammed 622	Reign of Charlemagne: 768 to 814	Caliphate of Cordoba: 929-1031	First Crusade, Fall of Jerusalem: 1099	Founding of the Order of the Temple: 1118	Genghis Khan founds the Mongol Empire: 1206	Travels of Marco Polo: 1271-1295	The Mongols rule China: 1279	Fall of Constantinople: 1453
Scientific or Technological Event	Building of Santa Sophia: Constantinople 532-537	Applications of industrial chemistry: Fireworks in China 605	Use of gunpowder in China: 9th century	History of Medicine by Avicena: 1012	Romanic Cathedral at Santiago de Compostela: 1075-1128	Founding of the University of Bologna: 1088	Chinese navigators use a magnetic compass: 1150	Averroes, advances in Medicin and Philosophy: 1126-1198	Gothic cathedral of Burgos: 1221-260	The Aztecs found Tenochtitlán on Lake Texcoco: 1325
2.- CHINESE MACHINES AND INVENTIONS										
		Beginning of the Tang Dynasty: 618	Beginning of the Sung Dynasty: 960-979	"Wu Jing Zong Yao" (1040)	"Meng Xi Bi Tan" (1086)	Su Song builds his astronomical clock: 1089	"Xin Yi Xiang Fa Yao" (1089)	"Keng Chih Thu" (1149)	Genghis Khan forms the Mongol Empire (1206-1229)	"Nong Shu" (1313)
3.- MECHANICAL ENGINEERING IN ANTIQUITY										
4.- MEDIEVAL MACHINES AND MECHANISMS										
		"Treatise on Ingenious Devices" by the Banu Musa brothers (S. IX)	"Elementary Treatise on the Art of astrology" and "The Moon Box" Al Biruni (975)	"De Rerum Naturis" by Hrabanus Maurus (1022)			"The Book on the Construction of Clocks and their Use" by Al-Saati Al-Kurasani (1203)	"Treatise on the Knowledge of Mechanisms" by Al-Jazari (1206)	"Travel Notebook" by Villard de Honnecourt (first half 13th C)	
5.- THE MACHINE RENAISSANCE										
										"De Ingensis" by Mariano di Jiacopo "Il Taccola" (1382-1458)
6.- MACHINES OF THE IBERIAN COLONIAL EMPIRES										
7.- THE MACHINERY OF THE INDUSTRIAL REVOLUTION										
8.-SOME THOUGHTS ON MACHINES										
	"Dark" period in the High Middle Ages, during which the main scientific-technological advances were produced in the East and in the regions under the influence/control of Islam									

TIME LINE	EARLY MODERN PERIOD (1453 - 1789)												
Historical Event	Discovery of America: Columbus 1492	The Ottoman Empire conquers Jerusalem: 1517	Spain invades the Aztec Empire: 1502-1520	Round-the-World Expedition: Magellan and Elcano from 1519 to 1522	Luther's translation of the Bible: 1522	Mayas conquered by the Spanish: 1524-1697	Fall of the last Inca resistance: 1572	The 30 Years War: 1618 to 1648	Consolidation of the Russian Empire with Peter I: 1689	Mongolia conquered by the Chinese: 1697	Spanish War of Succession: 1702-1713	Independence of the United States: 1783	French Revolution: 1789
Scientific or Technological Event	Printing Press: Gutenberg 1455	Leonardo da Vinci: 1452-1519	Heliocentrism: Nicholas Copernicus 1543	Pulmonary circulation of the blood: Miguel Servet 1553	Galileo Galilei: 1564-1642	Johannes Kepler's Laws: 1609	Descartes: Cartesian Method and Studies on Geometry 1637	Isaac Newton: 1642-1727	Gottfried Leibniz: 1646-1716	Newton's "Philosophiae Naturalis Principia Mathematica": 1687	Leonhardt Euler: 1707-1783	Steam Engine: James Watt 1769	Law of Conservation of Mass: Lavoisier 1789
2.- CHINESE MACHINES AND INVENTIONS													
	Beginning of the Ming Dynasty: 1368					"Cheng Chhüan Shu" (1628)	"Thien Kung Khau Wu" (1637)	End of the Ming Dynasty: 1644					
3.- MECHANICAL ENGINEERING IN ANTIQUITY													
4.- MEDIEVAL MACHINES AND MECHANISMS													
5.- THE MACHINE RENAISSANCE													
	"Trattato di architectura e machine" by Francesco Di Giorgio (1480)	Leonardo da Vinci's Codex (1480)	"Re Militari" by Roberto Valturio (1535)	"Theatrum Instrumentum et Machinarum" by Jacobus Bessonus (1578)	"De Re Metallica" by Georgius Agricola (1556)	"Le Diverse et Artificiose Machine" by Agostino Ramelli (1588)	"Novo teatro di machine et edificii" by Vittorio Zonca (1600)	"Theatri machinarum erster" by Henrich Zeising (1611)	"Kunstliche Abrís allerhand Wasser..." by Jacobus Strada (1617)	"Le Machine" by Giovanni Branca (1629)	"Theatrum Machinarum Novum" by Georg Böckler (1661)		
6.- MACHINES OF THE IBERIAN COLONIAL EMPIRES													
				"The Twenty-One Books of Devices and Machines" attributed to Juanelo Turriano (1570)		Francisco Lobato's Notes on Technology (1547-1585)		Patents for over fifty devices for Jerónimo de Ayanz y Beaumont (1606)				"The Royal Machine of the Segovia Mint" by Stephen Murray (1729)	
7.- THE MACHINERY OF THE INDUSTRIAL REVOLUTION													
												"The Miners Friend" by Thomas Savery (1689)	Thomas Newcomen invents a steam engine in 1712
8.-SOME THOUGHTS ON MACHINES													
				"Mechanicorum Liber" (1577) and "Le Mechanique" (1581) by Guidobaldo del Monte	"Theatrum Machinarum" (1588) by Agostino Ramelli					"Les Mecaniques" (1634) by Galileo Galilei		"Theatrum Machinarum Generale" and "Theatrum Machinarum Hidraulicarum" by Jacob Leupold (1724)	

TIME LINE	LATE MODERN AND CONTEMPORARY PERIODS (Since 1789)													
Historical Event	Napoleonic Wars: 1799-1815	Peninsular War: 1808-1813	Hispanoamerican War of Independence: 1809-1824	Communist Manifesto: Marx and Engels 1861	Unification of Italy: 1861	War of Secession: 1861-1865	Unification of Germany: 1871	Russian Revolution: 1905 (Red Sunday) and 1917	First World War: 1914-1918	Great Depression: 1929	Spanish Civil War: 1936-1939	Second World War: 1939-1945	Fall of the Berlin Wall: 1989	Attacks of 11 Sept: 2001
Scientific or Technological Event	First Vaccine (Smallpox): Edward Jenner 1796	First electric battery: Alessandro Volta 1799	First turbine built by Fourneyron: 1827	First train: (Liverpool - Manchester) 1830	Periodic Table of Elements: Mendeleiev 1869	Otto Engine: 1876	Invention of the motor car: Benz 1886	Diesel Engine: 1892	Invention of the airplane: Wright brothers 1906	Theory of Relativity: Einstein 1905 (Special) and 1915 (General)	Discovery of Penicillin: Alexander Fleming 1928	Invention of the computer (following Von Neumann architecture): EDSAC and Mark I 1949	First satellite: "Sputnik I" 1957	Complete sequence of the human genome: Weissenbach 2000
2.- CHINESE MACHINES AND INVENTIONS														
3.- MECHANICAL ENGINEERING IN ANTIQUITY														
4.- MEDIEVAL MACHINES AND MECHANISMS														
5.- THE MACHINE RENAISSANCE														
6.- MACHINES OF THE IBERIAN COLONIAL EMPIRES														
7.- THE MACHINERY OF THE INDUSTRIAL REVOLUTION														
	Watt's steam engine is built and Arkwright's "Water Frame" in 1769	John Wilkinson builds the first horizontal drill in 1775		"Portefeuille industrielle des machines, outils et appareils" (1841)		"Dissertation on the explanation and use of a new "machine for softening hemp and flax" by Salvá y Saponts (1874)		"Industry in 1874..." by José Alcover (1875)						
8.-SOME THOUGHTS ON MACHINES														
	"Essai sur la composition des Machines" by Betancourt and Lanz (1808)	"Traité Élémentaire des Machines" by Hachette (1811)	"Traité complet de mecánique appliquée aux arts" by Borgnis (1818)	"Principles of Mechanism" by Willis (1841)	"Principles of the Mechanics of Machinery and Engineering" by Weisbach (1848)	"Traité de cinématique" by Labouyale (1861)	"Die Bewegungs-Mechanismen" by Redtenbacher (1866)	"Kinematics of Machinery" by F. Reulaux (1875)	"Kinematische Modelle nach Prof Reuleaux" by Voight (1907)			Founding of IFTOMM (1969)	"Mechanisms in Modern Technology" by Artobolevski (1976)	"Machine Elements " by Niemann (1981)

References

1. G. Agrícola, *De Re Metallica* (Dover, NewYork, 1556) (Reprinted in 1950)
2. Anonymous. *Felipe II: los Ingenios y las Máquinas: Ingeniería y Obras Públicas en la época de Felipe II* (Exhibition: Engineering and public works at the time of Philip II, Royal Botanical Gardens, Instituto Cervantes, CSIC, 1998)
3. Anonymous. *Los Veintiún Libros de los Ingenios y de las Máquinas (The Twenty-One Books of Devices and Machines)* (Juanelo Turriano Foundation in 1997 in Spanish and English, in 7 volume edition, 1570)
4. A.R. Al Biruni, *Al-Tafhim-li-Awail Sina'at al-Tanjim* (10th Century Work on Mathematics and Astronomy)
5. M. Alcan, *Traité du travail de la laine cardée (Working Treatise on Carded Wool)*, 2 Vols. and 1 Atlas (Librairie Polytechnique Baudry, 1867)
6. J. Alcover Sallent, *La Industria en 1874: Revista de las Máquinas más perfeccionadas y de los Progresos realizados en los diversos ramos de la Industria según los datos recogidos en la Exposición Universal de Viena y de los adelantos hechos con posterioridad a dicho certamen.* A review of the Industry in 1874 (Manuel Tello Foundation and Printers, 1875)
7. T. Al-Din, *Al-Turuq al-samiyya fi al-alat al-ruhaniyya (The Sublime Methods of Spiritual Machines)* (1551)
8. M. Alheilig, *Traité des Machines à Vapeur: redigé conformement au programme du cours de machines à vapeur de l'École Centreale (A Treatise on Steam Engines)* (Gauthier-Villars et Fils, 1895)
9. A. Al-Yazari, *Kitab fi macrifat al-hiyal al-handasiyya (The Book of Knowledge of Ingenious Devices)*, 1206. Trans. D. Hill with annotations (D. Reidel, Dordrecht, 1974)
10. L. Allievi, *Cinematica della Biella Piana* (Francesco Giannini e Figli, Napoli, 1895)
11. L. Antoccia, *Leonardo, Arte e Scienza* (Giunti Editore, Firenze Italy, 2000)
12. I.I. Artobolevski, *Mechanisms in Modern Engineering Design*, 7 vols. (Mir Publishing, Moscow, 1976)
13. J. Badeau, *The Genius of Arab Civilization: Source of Renaissance* (MIT Press, Cambridge, 1978)
14. B. Musa, *Kitab Al-Hiyal (The Book of Ingenious Devices)* (9th century)
15. D. Barbaro, *I Dieci Libri dell'Architettura di Vitruvio, tradotti e commentati (The Ten Books of Vitruvius' Architecture, translated and commented)* (1556, Re-published by M. Bieber in 1961)
16. G. Basalla, *La Evolución de la Tecnología.* (RBA, D.L., 1994)
17. E. Bautista Paz, A computer-oriented method for calculus of Mechanisms. European Systems Engineering Symposium, 1965
18. E. Bautista Paz, J.L. Muñoz Sanz, P. Leal Wiña, J. Echávarri Otero, *Industrial Archaeology: from the XVII to the XXI Century. Reconstruction of Herschel's Telescope.* Proceedings of HMM2004, 2004
19. J. Bessonus, *Theatrum Instrumentum et Machinarum* (1578)

20. J. Beggs, *Mechanisms* (McGraw-Hill, New York, 1955)
21. M. Berg, *The Age of Manufactures, 1700-1820. Industry, Innovation and Work in Britain* (Routledge, London, 1985)
22. F.A. Borgnis, *Traité complet de mécanique appliquée aux arts (Complete Treatise on Mechanics Applied to the Arts)*, 9 vols. (Bachelier, Paris, 1818)
23. P. Bourguignon, *Cours de Cinematique Theorique et Appliquée II. Cinematique Appliquée* (Paris, 1906)
24. G. Böckler, *Theatrum Machinarum Novum* (1661)
25. G. Branca, *Le machine* (1629)
26. R. Bricard, *Leçons de Cinematique*, 2 vols. (Gauthier-Villars, Paris, 1927)
27. H.T. Brown, *Five Hundred and Seven Mechanical Movements* (Brown, Coombs & Co., New York, 1871)
28. L. Burmester, *Lehrbuch der Kinematik* (Felix Verlag, Leipzig, 1888)
29. E. Cano et al., *Moda, diseño y revolución. El nacimiento de la máquina de coser (Fashion, Design and Revolution. The Birth of the Sewing Machine)* (Mapfre Foundation, 2002)
30. J. Caro Baroja, *Tecnología Popular Española* (Círculo de Lectores, South Africa, 1996)
31. P. Casati, *Mechanica* (Lyons, 1684)
32. M. Ceccarelli, *The Challenges for Machine and Mechanism Design at the Beginning of the Third millennium as Viewed from the Past.* Proceedings of COBEM2001, 2001
33. M. Ceccarelli, *An Historical Perspective of Robotics Toward the Future.* Fuji Int. J. Robot. Mechatron. **13**(3), 299–313 (2001)
34. M. Ceccarelli, *El Renacimiento de las Máquinas: Primer Desarrollo de la Ingeniería Mecánica Moderna* (Congress on the History of the Legacy of Engineering, Las Palmas, 2006)
35. C. Cesariano, *Di Lucio Vitruvio Pollione de Architectura Libri Dece traducti de latino in vulgare, raffigurati, commentati* (1521)
36. T.G. Chondros, Deus-ex-Machina. International Conference on Ancient Greek Technology. Salonica, Greece, 2004
37. J. Dantín Cereceda, *La vida de las plantas (The Life of Plants)* (Espasa-Calpe, 1944)
38. P. Deane, *The First Industrial Revolution* (Cambridge University Press, Cambridge, 1998)
39. K.H. Decker, *Maschinenelemente Gestaltung und Berechnung* (Carl Hansen Verlag, Munich, 1971)
40. J.M. Díaz Rodríguez, *Molinos en Gran Canaria (Mills on Grand Canary Island)* (Caja Insular de Ahorros de Gran Canaria, 1988)
41. L. Euler, *Theoria Motus Corporum Solidorum seu Rigidorum* (Apud Bernuset, Delamolliere, Falque & Soc., 1765)
42. R. Franke, R. Oldenbourg, *Eine Vergleichende Schalt und Getriebelehre – NeuWege der Kinematik* (Munich, 1930)
43. F. Freudenstein, G.N. Sandor, *Synthesis of a Path generating Mechanism by a Programmed Digital Computer*. J. Eng. Indus. ASME. **81**B(2) (1959)
44. G. Galilei, *Mechanics, 1634* (Re-edited by F. Brunetti) (Turin, 1964)
45. J.A. García-Diego, *Los Relojes y Autómatas de Juanelo Turriano (Juanelo Turriano's Clocks and Automatons)* (Albatros, 1982)
46. N. García Tapia, *Pedro Juan de Lastanosa y Pseudo Juanelo Turriano.* Llull Magazine, vol. 10, N. 18–19 (1987)
47. N. García Tapia, *Técnica y poder en Castilla durante los siglos XVI y XVII (Technology and Power in Castille during the 16th and 17th Centuries)* (Consejería de Cultura y Bienestar Social, 1989)
48. N. García Tapia, *Patentes de Invención Españolas en el Siglo de Oro (Spanish Invention Patents in the Golden Age)* (Industrial Property Register, D.L., 1990)
49. N. García Tapia, *Un Inventor Navarro. Jerónimo de Ayanz y Beaumont* (Pamplona Department of Education and Culture, 2001)
50. N. García Tapia, J.A. García Diego, *Vida y Técnica en el Renacimiento: Manuscrito de Francisco Lobato, de Medina del Campo, en el siglo XVI. (Life and Technology in the*

Renaissance: Manuscript by Francisco Lobato, from Medina del Campo, in the 16th Century) (Valladolid: Publications Department, University, D.L., 1990)
51. N. García Tapia, J. Carrillo Castillo, *Tecnología e Imperio. Ingenios y leyendas del Siglo de Oro (Technology and Empire. Golden Age Devices and Legends)* (Turriano, Lastanosa, Herrera, Ayanz, Nivola, 2002)
52. F. di Giorgio, *Trattato di Architectura e Machine* (16th century)
53. Z. Gongliang, D. Du, *Wu Jing Zong Yao (Collection of the Most Important Military Techniques)* (1040)
54. M. Grübler, *Getriebelehre* (Springer, Berlin, 1917)
55. M. Gürtler, *Textil-industrie* (Walter de Gruyter & Co., 1928)
56. M. Hachette, *Traite Élémentaire des Machines* (Corby, 1811)
57. K. Hain, *Applied Kinematics* (McGraw-Hill, New York, 1967)
58. J. Halling, *Principles of Tribology* (MacMillan, New York, 1978)
59. R.S. Hartenberg, J. Denavit, *Kinematic Synthesis of Linkages* (McGraw-Hill, New York, 1964)
60. H. Hass, *Wie der Fisch zum Menschen wurde (How Fish Became Humans)* (Bertelsmann, Munich, 1979)
61. C.W. Ham, E.J. Crane, W.L. Rogers, *Mechanics of Machinery* (McGraw-Hill, New York, 1958)
62. D. Hill, A. Al-Hassan, *Islamic Technology: An Illustrated History* (Cambridge University Press, Cambridge, 1986)
63. D. Hill, *Studies in Medieval Islamic Technology: From Philo to al-Jazari: from Alexandria to Diyar Bakr* (Variorum Reprints, 1998)
64. V. de Honnecourt, *Notebook from the 13th Century* (Republished by Ed. Akal) (1991)
65. M. di Jacopo il Taccola, *De Ingensis* (Republished by L. Reichert in 1984) (1433)
66. B.W. Kennedy, *Kinematics of Machinery by Franz Reuleaux*, vol. 81, pp. 159–168 (MacMillan, London, 1876)
67. C. Kyeser, *Bellifortis* (1405)
68. A. Kircher, *Oedipus Aegyptiacus* (1654)
69. C-P.L. Laboulaye, *Traité de cinématique, ou théorie des mécanismes* (E. Lacroix, 1861)
70. A. de Lamadrid, A. de Corral, *Cinemática y Dinámica de Máquinas* (E.T.S.I. Industriales, 1963)
71. D. Landes, *Technological Progress and the Industrial Revolution* (Tecnos, D.L., 1979)
72. D. Laurenza, M. Taddei, E. Zanon, *Illustrated Atlas of Leonardo's Machines* (Giunti Editore, Milano, 2005)
73. J. de Lanz, A. de Betancourt, *Essai sur la composition des machines* (París, 1808)
74. J. Legazpi, *Ingenios de Madera (Wooden Devices)* (Caja de Ahorros de Asturias/Ministry of Agriculture, Fisheries and Food, 2001)
75. J. Leupold, *Theatrum Machinarum Generale* (Druckts Christoph Zunkel, 1724)
76. J. Leupold, *Teatrum Machinarum Hidraulicarum* (Druckts Christoph Zunkel, 1724)
77. A. de Liñán Vicente, *Entomología agroforestal (Farming and Forestry Entomology)* (Ediciones Agrotécnicas, D.L., 1998)
78. H. Maurus, *De Rerum Naturis* (1022)
79. M. Mauss, *Introducción a la Etnografía (Introduction to Ethnography)* (Istmo, 1967)
80. H. de Monantheuil, *Aristotelis Mechanica* (1599)
81. A. Montagu, *Les Premiers Âges de L'homme (The First Ages of Man)* (Seghers, 1957)
82. G. del Monte, *Mechanicorum Liber* (1577)
83. G. del Monte, *Le Mechanique* (1581)
84. G. Morís Menéndez-Valdés, *Ingenios hidráulicos históricos: Molinos, batanes y ferrerías (Historical Hydraulic Devices: Mills, Fulling Mills and Foundries)* (E.T.S.I, Gijón, 2001)
85. G. Murray, J.M. Izaga, J.M. Sole, *El Real Ingenio de la Moneda de Segovia: maravilla tecnológica del siglo XVI (The Royal Machine of the Segovia Mint: 16th Century Technological Marvell)* (Foundation Juanelo Turriano, 2006)
86. J. Needham, *Science and Civilisation in China* (Cambridge University Press, Cambridge, 1975)
87. B. Neroni, *Notes on Pumps* (1550)

88. G. Niemann, *Maschinenelemente*, 3 vols. (Springer-Verlag, Heidelberg, 1967)
89. J. Nieto, *Síntesis de Mecanismos (Synthesis of Mechanisms)* (AC, 1978)
90. R. Pareto, *Enciclopedia delle Arti e Industrie* (Unione Tipografico-Editrice, 1878–1898)
91. J. Persoz, *Essai des Matières Textiles: Méthodes et Appareils en usage (Essay on Textile Materials)* (Gauthier-Villars, 1899)
92. W. Plum, *Natural Sciences and Technology on the Road to the "Industrial Revolution"* (Friedrich-Ebert-Stiftung, 1974)
93. A. Ramelli, *Le Diverse et Artificiose Machine* (1588)
94. W. Rankine, *Manual of the Steam-engine and Other Prime Movers* (Charles Griffin, London, 1859)
95. F. Redtenbacher, *Die Bewegungs-Mechanismen* (Heidelberg, 1866)
96. A. Rees, *Clocks, Watches and Chronometers* (Charles Tuttle, Rutland, Vermont, 1819) (Reprinted in 1970)
97. F. Reuleaux, *Kinematics of Machinery* (MacMillan, London, 1876)
98. F. Reuleaux, *Lehrbuch der Kinematik*, 2 vols. (Braunschweig: F. Vieweg und sohn, 1876)
99. L. Reti, *El artificio de Juanelo en Toledo (Juanelo's Device in Toledo)* (Toledo Provincial Council, 1968)
100. J. Ribera, *Hiladura y tisaje (Spinning and Weaving)* (Est. Tip. de F. Nacente, 1887–1891)
101. G. Richard, *Traité des Machines-Outils (Machine Tool Treatise)* (Librairie Polytechnique Baudry et Cíe, 1895)
102. N. Rosenauer, A.H. Willis, *Kinematics of Mechanisms* (Dover, New York, 1967)
103. B. Roth, F. Freudenstein, Synthesis of path-generating mechanisms by numerical methods. ASME J. Engl. Indus. **81**, 1–7 (1963)
104. F. Salvá, *Disertación sobre la explicación y el uso de una nueva máquina para agramar cáñamos y limos (Dissertation on the Explanation and Use of a New Machine for Softening Hemp and Flax)* (Royal Printing House, Madrid, 1784)
105. G. Santi-Mazzini, *Hi-Tech '800. Machines* (Gribaudo, 2002)
106. E. Sauvage, *La Machine a Vapeur: Traité Général contenant la Théorie du travail de la Vapeur (The Steam Engine)* (Baudry et Cie, 1896)
107. T. Savery, *Miner's Friend* (S. Crouch, London 1689)
108. J.E. Shigley, L.D. Mitchell, *Mechanical Engineering Design* (McGraw-Hill, New York, 1983)
109. G. Schott, *Mechanicahydraulica-pneumatica* (Würtzburg, 1657)
110. J. Schröder, *Catalog of Reuleaux Models: Polytechnisches Arbeits-Institut. Illustrationen von Unterrichts-Modellen und Apparateni* (Polytechnisches Arbeits-Institut, 1899)
111. W. Schwoerbel, *Evolution, Strategie des Lebens* (Maier (Ravensburg), 1978)
112. M. Silva Suárez, *Técnica e ingeniería en España: El siglo de las luces (Technology and Engineering in Spain: The Century of Illumination)* (Institution "Fernando el Católico," Prensas Universitarias, 2004)
113. M. Silva Suárez, *Técnica e Ingeniería en España I. El Renacimiento. (Technology and Engineering in Spain I. The Renaissance)* (Real Academia de Ingeniería, 2004) (At the end are included 102 notes on engineers of the period and their Patents of Invention)
114. S. Song, *Xin Yi Xiang Fa Yao (Treatise about the Clock Tower)* (1089)
115. N. Stelliola, *De gli Elementi Machanici* (1597)
116. D. Stewart, *A Platform with Six Degrees of Freedom*. Proceedings of the Institute of Mechanical Engineering, London, vol. 180, pp. 371–386, 1965
117. J. Strada, *Kunstliche Abrisz allerhand Wasser- Wind- Rosz- und Handt Muhlen* (A Renaissance work on Machines, 1617)
118. S. Strandh, *Machines: An Illustrated History* (Artists House, London, 1979)
119. S. Strandh, *A History of the Machine* (Barnes & Noble, 1981)
120. T. al Din, *Al-Turuq al-saniyya fi al-alat al-ruhaniyya (The Sublime Methods of Spiritual Machines)* (1551)
121. N. Tartaglia, *Quesiti et Inventioni diverse* (1546)
122. R. Thurston, *A Manual of the Steam-Engine* (Wiley, New York, 1891)

123. R. Valturio, *Re Militari* (Apud Christianum Wechelum, 1535)
124. Various authors, *Publication industrielle des Machines, Outils et Appareils* (Librairie Technologique Armengaud Aine, 1841)
125. Various authors, *Portefeuille Economique des Machines* (Librarie polytechnique ch. Béranger, 1900)
126. Various authors, *Crónica de la técnica* (Plaza y Janés, 1993)
127. Various authors, *El oro y la plata de las Indias en la época de los Austrias (The Gold and Silver of the Indies in the Period of the Austrias)* (Fundación ICO, 1999)
128. Various authors, *International Symposium on the History of Machines and Mechanisms*, ed. by M. Ceccarelli (Kluwer, Dordrecht, 2000)
129. Various authors, *International Symposium on the History of Machines and Mechanisms*, ed. by M. Ceccarelli (Kluwer, Dordrecht, 2004)
130. Various authors, *Nature Magazine* **445**(7130) (2007)
131. Ch. Vigreux, *Turbines: Turbines Centripètes, Turbines Mixtes, dites Américaines, Roues Vives à Réaction (Pelton etc.)* (East Bernard, 1899)
132. L. da Vinci, *Atlantic Codex* (15th century)
133. L. da Vinci, *Madrid Codex* (1493)
134. L. da Vinci, *Leonardo's Notebook*, ed. by H. Anna Suh (Parragon, 2005)
135. P.M. Vitruvius, *De Architectura*, ed. by F. Giocondo (Verona, 1511) (Reprinted in 1513, 1522 and 1523)
136. G. Voight, *Kinematische Modelle nach Prof. Reuleaux* (Gustav Voigt Mechanische Werkstatt, 1907)
137. B. Woodcroft, *The Pneumatics of Hero of Alexandria* (Taylor Walton and Maberly, London, 1851)
138. J. Weisbach, *Principles of the Mechanics of Machinery and Engineering* (Lea & Blanchard, Philadelphia, 1848)
139. R. Willis, *Principles of Mechanism* (Longmans, Green, London, 1841), 2nd edn., 1870
140. H. Zeising, *Theatri Machinarum Erster* (1611)
141. W. Zeng, *Nong Shu* (Medieval Treatise on Agriculture, 1313)
142. V. Zonca, *Novo Teatro di Machine et Edificii per uarie et sicure operationi* (1607)

CPSIA information can be obtained at www.ICGtesting.com
Printed in the USA
LVOW071019220412

278635LV00003B/11/P

9 789048 125111